室内设计风格详解

MANUAL OF INDUSTRIAL INTERIOR DESIGN

工业风

凤凰空间·华南编辑部 编

江苏凤凰科学技术出版社

目 录

CONT

第 3 章 / Chapter 3

拒绝平庸的工业风格设计

The Unique Industrial Style Interior Design

居住空间
Residential Space

CONT

ENTS

↘ 办公空间
Office Space

第1章

Chapter 1

工业风格的形成和发展

The Formation and Development of Industrial Style

1 工业风格的形成

The Formation of Industrial Style

工业风格的历史可以追溯到 19 世纪末至 20 世纪初工业革命进入高速发展的年代，工业革命解决了困扰人类已久的生产原动力问题，机器的产生和使用大大地提高了生产效率，解放了人力，而工厂体制的建立，彻底将人类卷入了工业化的漩涡。工业风格也在工业革命强有力的推动下，逐步迈上了历史的舞台。

It could be dated back to the beginning of 19th to the end of 20th century when the Industrial Revolution has met its rapid development that the Industrial Styles came into existence. The revolution has solved the problem in manufacturing power that has perplexed the human beings for a long period. The creation and appliance of machinery has boost manufacturing efficiency, released manpower, and with the establishment of factory related mechanism, mankind set their foot in the industrial territory. And the industrial styles, driven by the Industrial Revolution, gradually took their place in history.

工业革命宣告了传统手工艺生产方式的没落，机械化、批量化的大生产促进了社会化大分工。工业风格自诞生起就与工业设计紧密联系到一起，没有工业设计的不断发展就没有工业风格的历史演化。工业设计集艺术、技术和商业于一体，是一门改善产品外观并提高产品价值的艺术。19 世纪后半叶，生活消费品大幅增加，这时的产

费城百年展览机械厅的内部，机械厅与临时建筑合为一体，由 Pettit 和 Wilson 在 1876 年设计，摄影：Centennial Photographic Co.
Interior View of Machinery Hall of the Centennial Exhibition in Philadelphia, Constructed as one of a Temporary Buildings, Designed by Pettit and Wilson at 1876, Photographer: Centennial Photographic Co.

品创意比历史上任何一个时期都更为举足轻重，因为消费者愿意为豪华、实用的产品付出更多。

The Industrial Revolution has put an end to the traditional handicraft production era, since mechanization and mass production has greatly accelerated the Great Social Division of Labor. The industrial styles related closely to the industrial design since its beginning, and without the constant development of which, there will hardly be any evolution in history of these styles. Combined with art, technology and commerce, the unity of industrial styles is a form of art that enhances value and appearance. In the latter half of the 19th century, consumer's products increased dramatically, which added importance to product innovation to a point that no other period of history can match, because consumers are willing to pay more for luxury and practical products.

工业设计建立在工业美学（Industrial Aesthetics）的基础之上。工业美学的理念必须通过工业设计来实现，因此也被称为“技术美学”和“商品美学”。工业美学主要体现在 4 个方面：功能美、技术美、材料美、形式美。工业美学追求锐利的线条、生硬的连接和强劲的互动，它是现代工业的机械制造、现代建筑的解构主义和机甲动漫的衍生物。它与传统的应用艺术不同，其审美来自于颠覆性技术与工业的发展。

The industrial design is based on the Industrial Aesthetics, the idea of which must be achieved through industrial design; thus it was also called The Aesthetics of Technology and The Aesthetics of Merchandise. There are 4 major elements that represent this aesthetics: completeness of functions, advanced technology, well-chosen materials, and elegant formality. It pursues sharp lines, angular connections and strong interaction, for which it was defined as a derivation of modern machinery production, architecture, deconstruction, and robotic comics and animations. Unlike other traditional apply arts, its beauty comes from overturning technology and industrial development.

工业美学的发展从时间上看，基本可以分为三个时期：第一个时期是自 18 世纪下半叶至 20 世纪初期，工业设计处于萌芽与探索的起源时期。1839 年的杂志《艺术联盟》首次出现了“工业设计”一词，此时的设计改革使传统的手工艺设计逐步向工业设计过渡，并探索出现代工业设计的发展道路，从而奠定了工业美学的基础。第二个时期是在第一次世界大战和第二次世界大战之间，是现代工业设计的形成与发展时期。工业美学已形成较系统的设计理论，并逐渐影响全世界范围的设计活动，工业风格依然更关注实用性的设计。第三个时期是在二战后，工业设计与工业生产以及科技发展联系异常紧密，工业美学多元化格局形成，工业风格也彻底从实用性中解脱出来，成为独立的建筑、室内设计风格。

From the perspective of time, the development of Industrial Aesthetics can be divided into 3 phases. The first is from the latter half of the 18th century to the beginning of the 20th century, when industrial design has just emerged and started its exploration. The phrase 'industrial design' first appeared in *The Art Union*, 1839, when the design revolution at the time promoted a transition from traditional handicraft design to industrial design, and found pathways for modern industrial design; hence laid down a foundation for the industrial aesthetics. The second phase is during the time between the World War I and World War II, when the modern industrial design took shape and was developed. More systematic theories of design have been developed, and the Industrial Aesthetics started to influence design activities globally, while still focusing on practicality. The third phrase is after the World War II, when industrial design, manufacturing, and the development of science and technology are bounded extraordinarily close together. As the pluralistic patterns of the Industrial Aesthetics have been formed, the industrial styles were completely released from practicality, and has become an independent style of architecture and interior design.

德意志制造联盟在推动工业美学运动发展当中发挥了重要的作用。作为德国官员的赫尔曼·穆特修斯在考察英国工业管理体系后提出“德国的工业设计应该反对继续模仿古旧的那些似是而非的形式”的主张。在他的倡导下，集合了银行家、企业家、商人、艺术家、建筑师、工程师以及政府官员的德意志制造联盟宣告成立，其章程提出“联盟的目标是与艺术、工业、手工艺通力合作，通过教育、宣传以及对有关问题非官方的表态来完善行业的工作”，最终德国设计师走出了艺术与手工艺运动的局限，奠定了工业设计的基础，并确保了工业风格发展方向。

The Deutscher Werkbund has contributed significantly in the promotion movement of the Industrial Aesthetics. As a German officer, Hermann Muthesius, brought forth the idea after inspecting the British industrial management system that the German industrial design should oppose continuous imitation of those antiquated and paradoxical forms. Under his call, bankers, entrepreneurships, businessmen, artists, architects, engineers, and government officers gathered to establish the Deutscher Werkbund, whose constitution being that "The goal of the unity is to cooperate closely with the art, the industry, and the handicraft, and to optimize works in the industries through education, propaganda, and unofficial declaration." The German designers eventually broke the limit of the art and handicraft movement, set the foundation of industrial design, and solidified the direction of industrial styles.

1919 年德国著名建筑大师瓦尔特·格罗皮乌斯在德国魏玛市设立“公立包豪斯学校”，后改称“设计学院”，但习惯上仍称为“包豪斯”。包豪斯是德语 Bauhaus 的译音，由德语 Hausbau（房屋建筑）一词倒置而成，所以 Bauhaus 从字面上理解就是“造房子”。包豪斯致力于结合美术和工艺的创新，打破了纯美术与实用美术对立的局面，奠定了工业设计的基本理念：废弃传统建筑形式以及与功能无关的装饰，主张形式服从功能，强调形式的单纯、简单、明晰、尊重结构自身的逻辑，促进标准化并考虑商业因素。包豪斯通过设计实践，实现了技术与艺术的真正统一，形成了设计中的理性主义原则，开创了面向现代工业设计的设计方法，填补了长期以来一直存在于现代艺术与技术、手工艺之间的鸿沟。

In 1919, the famous German designer Walter Gropius setup the Public School of Bauhaus later renamed as Design Academy at Weimar, Germany, but still called Bauhaus. The word 'Bauhaus' was created by putting backward the German word 'Hausbau' which means 'house-building', so literally, 'Bauhaus' means 'building houses'. Bauhaus worked towards the incorporation of innovation form art and techniques, broke the confrontation between pure arts and applied arts, established some fundamental conception of the industrial design: abandoning traditional forms of architecture and non-functional decoration; advocating submission

包豪斯
Bauhaus

of forms to functions; emphasizing pureness, simplicity, clearness and comprehensibility; respecting the logic of structure itself; and promoting standardization and commercial elements. By means of practice, Bauhaus has united technology and art together, formed the rationalistic principles of design, initiated methodologies oriented to modern industrial design, filled the gaps between modern art, technology, and handicraft that has long existed.

工业风格意味着革新工业美学特征，通常是对一个旧房屋或者废弃的旧厂房内部进行改造，将其脱胎换骨打造成一个充满现代设计感的居所。水泥砂浆是工业风格的标志性装饰材料，抛光处理后的地面和墙面，成了今天充满早期工业时代机械美的必要搭配。

Industrial style means to innovate for new elements of Industrial Aesthetics, which is usually some reform made to some old or abandoned factories and buildings from inside, to build these into a modern sense residence. Cement and mortar are the landmark decoration material of industrial style, with polished ground and wall surface, which has become the necessary collocation of mechanical beauty in the early industrial age.

20 世纪工业现代化运动的倡导者们以极大的热情接受了前辈们的成果，并将这种新思想的传播作为一种道义上的责任。勒·柯布西耶（Le Corbusier）在 1924 年出版的《走向新建筑》（*Vers une architecture*）一书中说：“建筑‘结构和装饰的形式’经过几个世纪的缓慢演变，在钢筋水泥的 50 年里为人类建筑史做出了巨大贡献之后，又经历一次彻底的变革”。通过现代化运动，工程建筑学获得了人们的一致认可。除此之外，人们对新技术和新材料不断认识的过程也通过不同方式表现出来。

法国普瓦西萨伏伊别墅，1928 —1930 年，勒 · 柯布西耶设计
Villa Savoye, Designed by Corbusier, From 1928 to 1930

The 20th century industry modernization advocate accepted, with great enthusiasm, the predecessors' achievements, and regarded the dissemination of this new idea as a moral responsibility. Le Corbusier said in his book, *Vers une architecture*, published in 1924: The structure and form of architecture, which has evolved over the centuries, has undergone a radical change after making a great contribution to the human history of architecture during the 50 years of concrete construction. Through the modernization movement,

萨伏伊别墅复原后的内部装饰
The Recovery Interior Design of Villa Savoye

architecture engineering was unanimously recognized by the people. In addition, the people's constant growing understanding of new technology and new materials are shown in different ways.

对机器和工业的崇拜在很大程度上奠定了意大利未来派和俄罗斯建设派的理论基础。密斯·凡·德·罗是一位奉行包豪斯的实用性原则的建筑师，他提出了"好的功能就是美的方式""功能绝对第一""少就是多"的现代主义设计原则，并通过教学影响了好几代现代设计师，对工业风格的形成具有重要的启发意义。他的设计通过大量使用玻璃来清楚地表露建筑结构。密斯·凡·德·罗 1929 年为巴塞罗那世界博览会设计的德国展馆，很有可能是第一座为充分发挥钢、混凝土等现代结构能力而建造的建筑。在这种结构中，墙体不再起承重作用，而可以自由地以各种材质和形式设置，使得室内空间可任意开敞和分隔，以满足不同的功能需求。

To a large extent, the worship of machines and industry laid the theoretical foundation of the Italian Futurist School and the Russian Construction School. Mies Van der Rohe is an architect who pursues practical of Bauhaus. He proposed modern design principles such as 'good function is the way of beauty', 'the function is of the absolute first', and 'less is more', and influenced generations of modern designers through teaching, which had significant impact on the formation of the industrial style. The extensive use of glass in his work clearly reveals the building's structure. Mies Van der Rohe's built for the German exhibition hall of the Barcelona World Expo at 1929 is likely to be the first construction for the purpose of making full use of steel, concrete and other modern structure ability. In this structure, the wall does not play the load-bearing role, but can be freely arranged in a variety of materials and forms, so that interior space can be arbitrarily opened and separated, in order to meet different functional requirements.

现代化运动相对于工业革命而言，对建筑设计和室内设计所产生的最明显的影响是"时代精神"的体现。在第二次世界大战中所取得的大量技术成果，使现代化的设计朝着工业性日趋突出的方向发展。而正是这个时候，建筑现代化开始显露出最初的分化产物——

重建的巴塞罗那世界博览会德国展馆
The Rebuilt Germany Pavilion in Barcelona World Expo

具有国际风格的作品。这种新形势的转变表现为建筑风格朝多元化发展，并使折衷主义和多元主义成为 20 世纪下半叶建筑的两个主要特征，这一转变不仅发生在建筑领域，而且也发生在室内设计领域。一方面，人们不断追求完美的心理是促成工业风格形成的根本原因之一；另一方面，对钢混结构时代和“工程建筑学”倡导的建筑实用性，二者的怀旧情绪所引发的复古风格也出现在战后英国的建筑设计中。

In the modernization movement-compared to the industrial revolution-it was the "spirit of the times' that had most noticeably influenced the architectural design and interior design. As great deal of technical achievements had been made in the World War II, modern design has been developing in an increasingly industrial direction. It was at this time that the modernization of architecture began to reveal the first differentiated product - an internationally styled work. This change in the new situation was represented by the diversified development of architecture that made eclecticism and pluralism two main characteristics of the second half of the twentieth century. This change not only took place in the field of architecture, but also in interior design. On one hand, people's intension of pursuing perfection is one of the fundamental reasons to the formation of industrial style. On the other hand, the age of steel and concrete and the theories of engineering structure have advocated practicability of architecture. Both had raised nostalgia that leads to the emerge of retro style in England.

这些思想于 20 世纪 50 年代末期在摄政街工业大学和伦敦建筑协会产生，主要倡导者为雷纳·本汉姆和亚奇格莱姆，他们曾经在很大程度上刺激了工业风格建筑师的创作灵感。这些建筑师们通过对一些经典理论学说的广泛宣传，将极为普通的建筑逐步发展为后来成为业内主流的工业风格建筑。这些理论学说的主要特点是提倡建筑的对称形式、散件的整体组装、预制构件的装配、航空航天材料的使用以及建筑师的环境保护意识。

These thoughts came from the Regent Street Polytechnic and the London Association of Architecture in late 1950s. The main

重建的巴塞罗那世界博览会德国展馆

The Rebuilt Germany Pavilion in Barcelona World Expo

advocates Reyner Banham and Archigram have stimulated the industry style architect's inspiration to a great extent. Through extensive publicity of some classic theories, the architects gradually developed the very ordinary buildings into the mainstream industrial style architecture in the industry. The main feature of these theories is promoting the symmetry of form, assembly of parts, prefabricated assembly, aerospace materials, and the architect's awareness of environmental protection.

早在 20 世纪的上半叶，已经有少数设计师开始提倡工业化的空间，而他们所强调的主要是实用性方面。美国设计师理查德·巴克敏斯特·富勒可谓是早期工业风格的先驱者，他因发展球体网架结构而闻名于世。这种结构之后一直被广泛采用，包括建造于 1967 年的蒙特利尔世界博览会的美国馆。它是一个用金属网架构成的大圆球，展品和活动都安排在球体内部。巨大的穹顶由塑料板封闭，光线透入其中，并由机械控制整体的明暗。室内展览空间设在通过自动扶梯可以到达的平台，围合的结构形成一个位于建筑上空的独立薄膜，造就了既美观又具有戏剧性的效果。

As early as the first half of the twentieth century, a small number of designers had begun to promote industrialization of space, and their emphasis was mainly on practicality. American designer Richard Buckminster Fuller is a pioneer in the early industrial styles. He has become famous in the world owing to his development of balled grid structure that has been widely used, including the Montreal International Expo built in 1967. It is a big sphere with a metal grid, inside which exhibits and activities are arranged. The great dome is enclosed by plastic plates that allow light penetration, and by machines, the overall shading is controlled. The interior exhibition space is built on platforms connected by escalators. The enclosed structure forms an independent film over the building, creating both beautiful and dramatic effect.

蒙特利尔世界博览会美国馆
America Pavilion in Montreal World Expo

2 早期工业风格
The Early Industrial Style

至 20 世纪 60 年代，早期工业风格即高技派风格（又称重技派）开始流行，这种风格是指以“机器美学”为原则，反对传统的审美观念，强调设计作为信息的媒介具有的交际功能，在建筑以及室内设计中采用新技术，突出当代工业技术成就。以金属、玻璃、塑胶、合成树脂等工业材料为主要材料表现建筑与空间。在室内暴露梁板、网架等结构，以及风管、线缆等各种设备，强调工艺技术和时代感。例如，把自动扶梯的下部装置明露。

The early industrial styles, i.e. the High-tech（also known as the Structural Expressionism） has become popular in the 1960s. It persists in the principle of "Machine Aesthetic", opposes the traditional aesthetic view, and emphasizes the communicative functions of design that acts a medium of information. It has also adopted new technologies for architectures and interior decoration, which highlights the achievements in the area of modern industry and technology. Metal, glass, plastic, and synthetic resin are used as major types construction materials in order to present the architectures and their space. Frameworks such as beam slabs and grid, or pipelines such as air duct and cables are exposed to emphasize technology and modernity, for instance, interior devices of an escalator are exposed.

高技派总是用透明的玻璃、半透明的金属网、金属构件把建筑的组成部分，比如：梁、柱、板、楼梯、管道等组合在空间中，并尽可能让它们运动（电梯、自动扶梯）。室内局部或者管道常常涂上红、绿、黄、蓝等鲜艳的颜色，增强室内的现代感。

The High-tech usually integrates construction components such as beams, pillars, walls, stairs, and pipes with highly transparent glass, translucence metal mesh, and metal parts, and tries to make them move as much as possible（e.g. lifts and escalators）. Parts of the interior space or the pipes are usually painted in red, green, yellow, or blue to enhance interior sense of present.

早期工业风格的本质是把现代设计中的技术因素加以提炼、夸张，形成一种符号的效果，由此获得一种新的美学价值和意义。在精神层面上，设计师们企图颠覆当时主流的、较为暖调的北欧式现代主义，利用冷峻、阳刚的气质显现出新潮感与时髦感。但同时，它的开放型、工业化和科技感，仍然是现代主义的延续。

The essence of the early industrial styles is its extraction and exaggeration on technical elements in modern designing in order to create a symbolic effect that produces a new sense of aesthetic values. Spiritually, designers attempted to overturn the Nordic Modernism, which featured in warmth and was dominating at the time, with coldness and masculine that demonstrate a sense of fashion. At the same time, its openness, industrialization, and sense of technology are still descendants of the Modernism.

早期工业风格的兴起与经济发展密切相关，20 世纪 60 年代商业结构和办公室剧增，对工业设计产品需求量加大，如办公室、室内装饰和办公用品等。对这些场合来说，有必要体现出商业界的秩序与效率，因此设计应带有冷漠、正规、中性的外观特征。现代主义在居家环境中体现出来的非正规、人情味和轻巧、灵活等特点已不适于商业办公的要求，从而促进了早期工业风格的兴盛。

The early industrial styles thrived closely with economic development. In 1960s, number of commercial architectures and offices has surged, which brought about greater demands on design for industrial products; for instance, offices, interior decoration, and stationery. For those locations, it is necessary to present the order and efficiency of business communities, designing for whose appearance should hence be of formality, coldness, and neutrality. That is to say, the informality, friendliness, swiftness, and lightness represented by the modern home environments no longer meet the needs of business, which then led to a rise of the early industrial styles.

带动早期工业风格室内设计风潮的是理查德·罗杰斯、诺曼·福斯特、伦佐·皮亚诺等建筑师以及美国室内设计师约瑟夫·保尔·乌尔素等。他们设计的作品看起来不像是一般的公共建筑，反而像精致化的工业厂房。这种建筑既新奇又功能齐全，因此形成一种极具科技感的新美学，在这种美学的启发下，一些设计师开始以工厂、办公室、医院、实验室、交通工具，甚至军事设施等较常用的冷硬建筑空间来建造住宅和商用空间。

Architects like Richard Rogers, Norman Foster, and Renzo Piano, and American interior designer as Joseph Paul D'urso has promoted a trend of the early industrial styles in the areas of indoor design. Their works look nothing like conventional public buildings, but resemble the refined industrial factory buildings. These types of constructions are both interesting and with complete functions, thereby creating a new kind of aesthetics that conveys strong sense of technology, and inspired by that, some designers started to put up residence houses and commercial spaces by cold and tough materials that are often seen in factories, offices, hospitals, labs, transportations, and even military facilities.

而最著名的早期工业风格建筑非巴黎蓬皮杜国家艺术文化中心莫属。它由意大利设计师皮亚诺与英国设计师理查德·罗杰斯共同合作设计。这座巨大的多层建筑将其结构暴露在外，并涂上鲜亮的工业标志性色彩，看上去像一座化工厂房。不同功能的管道用不同颜色进行表示：红色代表交通系统、绿色代表供水系统、蓝色代表空

巴黎蓬皮杜国家艺术文化中心外部自动扶梯
Centre National d art et de Culture Georges Pompidou External Escalator

巴黎蓬皮杜国家艺术文化中心外部彩色管道

Colorful Pipes Outside Centre National d art et de Culture Georges Pompidou

调系统、黄色代表供电系统、白色代表自来水管道。西边的建筑暗示了正在施工的建筑脚手架，而东边部分则宛如炼油厂或化工厂的管道。内部设有自动电梯，交叉贯穿于广场上空的巨大玻璃管道之间，使观众就像输油管道里的液体一样，流动于各大区间。

Designed cooperatively by Piano, an Italian designer, and Richard Rogers from Britain, the Centre National d art et de Culture Georges Pompidou has been counted as the most famous architecture featuring the these styles. With its structure exposed, the multistoried building was painted with bright industrial signal colors that makes it look like a chemical plant, with its pipes in different colors respectively by the functions-red representing the traffic system, green for the water supply, blue for the air conditioning system, yellow for the electric system, and white for the waterworks. The west part of the building implies a scaffold at a construction site, while the east part resembles the pipes of an oil refinery or a chemical factory. The huge glass pipe stretches over the square was installed with an escalator within, visitors on which travel are "flowing" like liquid in an oil pipe between each sector.

不同于外观的纷繁复杂，其内部空间却很简单。每层除了两侧 12 米宽用于布置自动电梯以及各种管道的空间外，其余将近两个足球场大的空间内没有一根柱子，也没有固定的墙面。通过家具、屏风或者预制构件在空间中作灵活的隔断分割，为艺术展览和表演空间，提供了灵活多变的空间环境。针对内部环境的独特设计，设计师这样解释："这幢房屋既是一个灵活的容器，又是一个动态的交流中心。自由和变动的情况是房屋的艺术表现。"

Unlike its apparent complexity, the building has rather simple interior arrangement. Except for the 12-metre wide space on each end of the hall, which is for the installation of escalators and various pipes, the huge space big enough to fit in tow football fields relies on neither any pillar nor any fixed wall. Furniture, screens, or prefabricated parts are used to achieve partition that provides flexible space arrangements for art exhibitions

巴黎蓬皮杜国家艺术文化中心室内空间

Interior Space of Centre National d art et de Culture Georges Pompidou and performances. Regarding the special interior designing, the designer explains: The construction is a flexible container as well as a dynamic communication centre. Freedom and changing situation are artistic presentation of architecture.

在巴黎蓬皮杜国家艺术文化中心工程结束后，伦佐·皮亚诺设计了位于美国德克萨斯州休斯敦的梅尼尔私人收藏博物馆，博物馆的外观颇具工业风格特色，灰色的松木板砌出模块化的长方形外墙，最引人注目的是吊挂于钢结构上，呈曲面状的预制混凝土"扇叶"反光板。曲面造型的反光板，是皮亚诺与结构大师彼得·莱斯合作的创新发明，将自然的光源漫反射于展示厅内，凸显了结构与科技的美感。

As the project of Centre National d art et de Culture Georges Pompidou came to an end, so as the cooperation of the two designer teams. The Menil Collection Gallery located in Huston, Texas, U.S.A. is Renzo Piano's creation, the appearance of whose exterior wall fairly features an industrial sense. The modular rectangle walls were built with gray pine boards, and the most noticeable thing is the prefabricated concrete curved "fans" reflectors. The curved mold reflector is an innovation brought by the cooperation of Piano and Peter Rice, which scatters natural light into the exhibition hall thereby highlighting the beauty of structure and technology.

梅尼尔私人收藏博物馆
Menil Collection Gallery

梅尼尔私人收藏博物馆预制混凝土"扇叶"反光板，可将太阳光拦截并反射到内部空间
Prefabricated Concrete Fan Blade' Reflectors that Catch and Reflect the Light to the Inside

梅尼尔私人收藏博物馆内部空间，摄影：D Jules Gianakos
Interior Space of Menil Collection Gallery, Photographer: D Jules Gianakos

梅尼尔私人收藏博物馆室外雕塑，摄影：D Jules Gianakos
The Sculpture Outside the Gallary, Photographer: D Jules Gianakos

位于伦敦的劳埃德大厦则是理查德·罗杰斯的另一个重要代表作品。罗杰斯在这个设计上，更加夸张地使用高科技元素，建筑外由两层钢化玻璃幕墙与不锈钢外装修构架组成，大厦内部楼板均架在钢筋混凝土井字形格架上，由巨大的圆柱支撑，柱内为钢筋混凝土结构，成排的自动扶梯设置在中央空间较为底层的范围内。建筑构件也遵循一定的模数设计，反映了建筑高科技化的新特点，不断暴露结构、大量使用不锈钢、铝材和其他合金材料构件，使整个建筑闪闪发光。这个像科学幻想一般的建筑，比他过去的巴黎蓬皮杜国家艺术文化中心更夸张、更突出，也使得“高技派”风格更为成熟。

The Lloyd's Building in London is another important representative of Richard Rogers' s works. In this design, Rogers make more exaggerated use of high-tech features. The outer layers are composed of two glass curtain walls and two exterior decoration frames of stainless steel. All floors inside the building are supported by the reinforced concrete grid, and huge pillars that are also reinforced concrete inside. Escalators are set abreast at the lower section of the centre space. Modulus design is adapted to the components as well, representing a high-tech trend featuring structure exposure, intensive use of stainless steel, aluminum, and other alloy components, hence creating a shimmering effect among the entire building. This sci-fi like architecture is even more exaggerated and magnificent than the Centre National d art et de Culture Georges Pompidou, which drove the High-tech to higher maturity.

劳埃德大厦
Lloyd's Building

劳埃德大厦内部
Interior Space of Lloyd's Building

中国香港汇丰银行大厦
HSBC Main Building in Hongkong

诺曼·福斯特被誉为高技派的代表人物之一。他设计的中国香港汇丰银行大厦不仅从空间的使用功能出发更极力追求表现技术美的特征。从外观结构上来看，大楼外形上显著暴露出的钢柱和钢桁架成为立面的主角，大厦处处显示现代技术的成就。刻意暴露的支撑结构，使银行的外观看来更像是一座高高耸起的加工厂，但精确的节点、极富韵律感的支架和价格昂贵的材料，无不给人秩序森严的印象，适当地体现了建筑作为银行所应具备的特性。

Norman Foster was honored as one of the representatives of the High-tech. In his work, HSBC Main Building, he not only proceeded from the use of space, but also pushed the beauty of technology to the greatest extent. From the perspective of appearance, its vertical face is attached with two dazzling steel pillars and steel truss that are, in the large extent, purposely exposed, exhibiting modern technological accomplishments. The deliberately shown supporting brings the bank to a resemblance to a highly raised processing factory. However, the accurately and rhythmically aligned joints and frames, and the costly material, are giving off impression of strictness and order that properly expresses the character a bank should possess.

詹姆斯·斯特林被认为是具有高技派倾向的英国建筑师。英国莱斯特大学工程馆以其玻璃办公塔楼、包含了演讲厅的楔形连接块和轮船漏斗形通风设备而吸引了广泛的注意。有一片大而矮的连接区域用于放置工厂设备。室内空间也表达了室外的机械特性，暴露的结构暗示了建筑与工程有关的角色定位。

James Stiring is known as a British architect with High-tech school tendency. The Engineering Building at Leicester University and its vitreous office tower draw much attention by the wedge-shape connection next to the lecture hall and the trumpet-shaped vents. There is a large and flat area for factory equipment and the interior space also displays the mechanical features of the outside. The exposure of structure indicates its architecture and engineering concerning personality.

莱斯特大学工程馆，由詹姆斯·斯特林设计
The Engineering Building at Leicester University, Designed by James Stirling

詹姆斯·斯特林后期最为重要的作品是德国斯图加特新国立美术馆（加建部分），在这里他摆脱了单纯的技术追求而转向更为大胆的方向。回廊空间环绕着圆形庭院布置，其中大理石墙、雕塑和塔司干柱式入口参考了传统的建筑风格，暗示了与过去艺术和建筑之间复杂的关系。展廊空间在形式和色彩上是受限的，而入口门厅、商店、人行空间和餐馆则采用了明亮饱和的色彩。涂有鲜艳颜色的换气管，带有构成主义痕迹和高技派味道的入口雨篷、素混凝土排水口、粗大的管状扶手等细部，虽然保持了高技派设计的准确性，但它显示设计师已经转向后现代的创作方法。

德国斯图加特新国立美术馆入口

The Entrance of New National Gallery of Stuttgart, Germany

The most important James Stiring's later work is the New National Gallery of Stuttgart, Germany （addition section）, for which he abandoned the pursuing of pure technology and turned to bolder directions. Corridors are arranged surrounding the circular courtyard, within which there are marble walls, sculptures and Tuscan column entrances that are built by referring to traditional styles, implying complex relationships in the past between art and architecture. The corridors are limited in the aspects of its forms and colors, while in the entrance hall, shops, restaurants and visiting areas, bright colors were applied. Although the High-tech's accuracy had been maintained, it has shown that the designer had shifted to post-modernity in designing methodologies. These changes can be presented by details of the brightly colored vent-pipes, the entrance rain shed in constructivism and High-tech style, plain concrete outfalls, and pipe-shape handrails.

德国斯图加特新国立美术馆

New National Gallery of Stuttgart, Germany

1987年建于法国巴黎的阿拉伯世界研究中心，是当代法国建筑师让·努维尔富有代表性的作品。此建筑无论室内室外都散发着工业风格的魅力：半通透、半反射玻璃给建筑带来一种飘忽不定的形象；北面弧形的墙面恰到好处地勾勒出塞纳河的走向；尖锐的弧形楔状体与长方体之间仿佛可以彼此滑动，呈现出动态的力的平衡。整个设计中，位于南立面的玻璃窗最引人注目，窗户上安装了类似相机快门结构的紧密的机械装置，由计算机控制采光量。大大小小的采光口组成了一个个绝妙的阿拉伯风格的图案，设计师运用现代技术创造性地表达了传统文化的内涵。

Finished in 1987, Paris, France, The Arab World Institute is a representative work of Jean Nouvel, contemporary French architecture. The building gives of charms of industrial style from both outside and inside: the semi opaque and half reflecting glass put uncertainty to the building; the north curved wall follows precisely the curse of the Seine; the sharp curved wedge body and the cuboids seemingly slides pass each other, presenting a dynamic balance of force. The most eye-catching element is the windows at the south vertical face. Motifs that were installed on the window are actually photo-sensitive motor-controlled apertures. The mount of light entering the building is controlled by computers. The combination of shutters forms pictures and pictures of marvelous Arabic patterns, expressing the traditional culture by modern technology in a creative way.

阿拉伯世界研究中心
The Arab World Institute

阿拉伯世界研究中心采光细部
The Detail of Lighting of The Arab World Institute

工业美学对住宅建筑领域的影响越发深远，设计师在单一家庭住宅中进行了大量试验性的设计，这些住宅主要由建筑师本人、其家人或者朋友居住。由皮埃尔·沙雷奥和伯纳德·比诺维于 1932 年设计的“玻璃之家”巴黎韦尔别墅是第一个将工业美学运用于住宅设计的例证。如遥控卷闸门、可调式玻璃窗、可收式楼梯、旋转式柜子以及金属书架，这些仅仅是建筑内部装饰中的一小部分。

Industrial Aesthetics affects housing architecture in an increasing way. Designers do large amount of experiments within houses, most of which are lived by the architectures themselves, their family members or friends. Designed by Pierre Chareau and Bernard Bijvoet in 1932, the Maison de Verre at Paris (The House of Glass) is the first example of applying industrial aesthetics to housing architecture. This can be exemplified by a small part of the decoration inside the building: the remote-controlled roller gates, adjustable glass windows, retractable stairs, rotatable cabinets and metal bookshelf, etc.

工业风格与极简风格相伴相生，相互影响深刻。韦尔别墅设计师运用了“少即是多”的设计理念。住宅本身是一个巨型的玻璃盒子，直径 3 米的红柱筒内包含着壁炉和浴室，将内部空间等分为三部分。建筑细节处理十分讲究，平滑光亮的钢构件尽可能贴近玻璃的内表面，以减少阴影和最大限度地增加透明感和反射效应。

The industrial style and minimalism affect and thrive with each other deeply. Designers of the Maison de Verre applied the idea of 'less is more', constructing the house into a huge glass box, with a red column casing wrapping the stove and bathroom, which trisects the interior space. The details are dealt with elegantly, with its smooth and shiny steel parts approaching the inner face of the glass to the largest extent so as to reduce shadow and maximize transparency and reflection.

↑ 巴黎韦尔别墅书房
The Study of Maison de Verre at Paris

巴黎韦尔别墅玻璃窗
Glass Windows of Maison de Verre at Paris

金属书架
Metallic Bookshelf

巴黎韦尔别墅楼梯
The Stair of Maison de Verre at Paris

查尔斯·埃姆斯在住宅与家具的设计上也展现了早期工业风格的特质。埃姆斯住宅使用暴露的开敞式网架支撑着屋顶，外墙用玻璃和实体板做成，采用了标准工业制作的窗户和结构构件的组合，常被用以证明以技术为基础的设计方法能够产生独具美感的空间。1946年以后，埃姆斯的家具设计也开始广为流行，埃姆斯椅更是受到了追捧。早期工业风格把工业环境中的技术特征引入到家庭产品和住宅设计中，自公共空间引入到了高度私密的个人空间。

Charles Eames' design in houses and furniture also exhibits the traits of the early industrial styles. For the Eames House, open grid structures are used as supporting of the ceiling. Its exterior walls are combined by glass and boards, and windows are made of components in accordance with industrial standards. That is usually meant to proof that designing based on technology is able to create unique spatial beauty. Since 1946, Eames' furniture started to become popular, especially the Eames lounge chair. The early industrial styles had brought technological features into home product and house design, and from the public space into the more private personal space.

埃姆斯住宅
Eames House

埃姆斯住宅内部
The Interior Space of Eames House

埃姆斯椅
Eames Lounge Chair

工业风格为了强调空间的工业感会刻意保留并利用那些曾经属于工厂车间的材料和设备，比如钢材、生铁、水泥和砖块等。有时候旧厂房内的暖气管道、管道灯具或者空调设备都会被小心地保留下来。但是由于过度重视技术和工业的表现，反而将装饰压到了最低限度，显得冷漠而缺乏人情。因此一些设计师也致力于创造出更富有表现力和更有趣味的设计语言来取代纯技术的体现，把“工业技术”与“艺术情趣”结合起来，于是波普艺术风格对工业风格产生了重要的影响。

The industrial style emphasizes the sense of industry among space by deliberately retaining and using materials and equipments that once belong to the factories, such as steel, pig iron, cement and bricks. Sometimes, heating pipes, pipe lights or air conditioners of an old factory building would be carefully preserved. However, due to the overly emphasizing presenting technology and industry, decoration had been minimized, which consequently made it lack of warmth and human kindness. Therefore, some designers embarked on creating designing languages that are more expressive and interesting to replace pure technological expression by combining 'industrial technology' and 'artistic sentiment'. Hence the Pop style has influenced the industrial style significantly.

波普（Pop）一词是流行艺术（Popular Art）的简称，原意为流行的、大众的。波普风格的产生与战后日渐形成的西方“丰裕社会”、青少年消费市场的需求有密切的关联，体现出一种叛逆的设计立场。波普风格打破了工业风格局限于现代国际主义风格的过于严肃、冷漠、单一的面貌，代之以诙谐、富于人性和多元化的设计。设计师在室内、日用品、家具、服饰和平面设计等方面的设计上，进行了大胆的探索和创新，表现出前所未有的形式：夸张、奇异、富于想象力的造型；色彩单纯、鲜艳；材料多选用塑料或廉价的纤维板、陶瓷等。

Pop, being the abbreviation of 'Popular Art', originally means popularity and publicity. The Pop style related closely with 'The Affluent Society' after the war, and the need of youngster consumers' market, hence reflecting a rebellious standpoint. The Pop style has broken up the solemnity, coldness, and dullness of the modern internationalism, which had limited the industrial style. Instead, it replaced that with humor, personality, and pluralism. The designers had explored and innovated boldly in the designing areas like interior designing, commodity, furniture, clothing, and graphic design in ways unprecedented: exaggerated, strange, and imaginative modeling; simple and bright coloring; and selections for material being mostly plastic and low-price fiber board or ceramic.

塞努塔公司设计的布娄充气沙发和皮尔罗·加提设计的萨库袋子沙发等都是波普家具的代表作品。前者是由 PVC 塑料制成的透明扶手椅，后者则是一个装满泡沫颗粒的“大口袋”，可根据人体任意成型，充满趣味。

The Blow （a kind of inflatable sofa） designed by Zanotta and the Sacco by Piero Gatti are typical products of Pop furniture.

萨库袋子沙发，由皮尔罗 · 加提在 1968 年设计，目前由塞努塔生产。表面材料含有高阻膨胀苯乙烯颗粒
Sacco Chair, Designed by Piero Gatti, Teodoro in 1968 and Currently Produced by Zanotta. Envelope Containing Highly Resistant Expanded Polystyrene Pellets

The former one is an transparent arm chair made by PVC, and the latter is an interesting large 'sack' filled with plastic foam particles that can be re-shaped to fit the human body.

波普风格的流行让工业风格开始注重减弱过于单调和冰冷的工业感，用一些柔和的材料（如纺织品和木材）与之搭配。同时为了强化个性和文化特征，墙面往往会被涂刷成醒目的色彩，颜色丰富且装饰感强，波普艺术作品也成为室内墙面最好的装饰品。

The popularity of the Pop drove industrial style towards reducing over-simplicity and coldness. Soft materials such as fabric and wood are applied to the combination. At the same time, to strengthen individualism and cultural traits, the walls are usually painted in bright colors. Being colorful and decorative, Pop style art works had also become the best adornment for interior wall faces.

早期工业风格建筑更加倾向于张扬地展示驾驭现代技术手段的能力，以极富工业特色的形象出现在城市之中。但随着人们对文化传统和历史给予更多的关注，设计师们也在思考如何运用有机的手法结合新科技以传达设计的理念，不断修正技术崇拜和技术至上的设计方向。

The constructions of the early industrial styles tend to favor the ability to control modern technological tools, and present themselves in the cities by images full of industrial characteristic. However, as people gradually pay more attention to cultural traditions and history, designers are contemplating how to convey their ideas by more organized methods and new science and technology, and constantly adjusting their pathway on worship of technology and technology orienting.

对于城市文脉的关注，使得工业风格建筑不再完全以自我为中心，从“城市中表演”到“为城市而存在”。这一转变无疑是十分深刻的，它使工业风格超越了一种简单的外在风格而进入更深层次的境界。法国加里艺术中心的出现完美体现出这一思想的转变。加里艺术中心位于法国尼斯城最著名的梅森卡里神庙旧址之上，这座历史悠久的神庙毁于 1952 年的一场大火，只余下建于 1827 年的恢宏的新古典主义风格入口。改造后的加里艺术中心展现了工业风格鲜明的特色：选用新颖、现代的材料，使用铝合金百叶天棚遮挡夏季的阳光，将柔和的漫反射光引入室内庭院，形成一个梦幻世界。更为特别的是为了使地下空间也获得自然光，室内的地板和楼梯踏板均采用透明的强化玻璃材质。而为了和梅森卡里神庙取得联系，艺术中心坐落于具有一定高度的石头基座上，前部是敞开的柱廊，建筑将近一半的空间位于地下，使得整体虽然规模宏大，但仍然显得低调细腻，与古老的街区融为一体，展现了建筑师对历史传统与现代建筑调和的功力。艺术中心的入口立面在虚实关系上恰好与对应的神庙侧面有着奇妙的相似。这一点被建筑师敏锐地抓住了，并认真地推敲形成了他们之间对应一致的比例关系，现代建筑与传统建筑进行着无声的对话。

加里艺术中心与梅森卡里神庙
Carr é e d'Art and Mason Cary Temple

加里艺术中心内部
The Interior Space of Carr é e d'Art

The focus on city context had made industrial style became less self-centered, which has shifted from 'perform in the city' to 'exist for the cities'. This transition is undoubtedly very profound, by which the industrial style has surpassed a kind of style simply of its appearance and rise to a higher level. The Carrée d'Art in France perfectly illustrated this shift as it appeared. It is located on the site of the Nice-famous Mason Cary Temple. In 1952, the historic building was burnt down, with only the fascinating neoclassical gate built in 1827 that remained. After reconstruction, the art center is displaying noticeable industrialism: modern materials are used. The aluminum shutter ceiling now blocks the summer sunlight, and allows it to the inside courtyard by diffused reflection, forming a dreamlike world. What's more special is that in order to shed light into the underground space, the floors and stair treads inside the building are all made of transparent reinforced glass. To create connection with the temple, the art center is situated on a stone base of certain height. The front of the center is an open colonnade, and nearly a half is below the ground. This has connected the building with the old streets, and the magnificent building itself, is still modestly exquisite, showing the architects' amazing skills that coordinate historical traditions and modern construction. The relationship between blankness and actuality among the entrance's vertical face just corresponds in an interesting way with one side of the temple. This had been grasped by the architects' perceptibility and through some serious contemplating, they managed to attain a corresponding ratio that brought about a silent conversation between the modern and traditional architectures.

德国国会大厦建于 1884 年，数次战争使它变得残缺破落。为了改变这一状况，诺曼·福斯特以建筑最初形式为蓝本，对国会大厦进行了修建，使之成为一座现代化的议会办公大楼。福斯特在国会大厦的顶层增加了一个巨大的玻璃穹顶，这个穹顶不仅与原建筑紧密结合，还将现代科技和生态原理运用到建筑之中。穹顶以玻璃和金属为主要材料，通过透明的穹顶和倒椎体的反射，将日光反射到议会大厅。白天可随日照方向自动调整方位的遮阳板将阳光引入大厅；晚上遮阳板摇身一变将室内灯光向外反射，使玻璃穹顶成为一座发光塔，为柏林的夜晚带来独特的景观。

The Reichstag was completed in 1884, and has been damaged during the many wars. To make a change to the situation, Norman Foster, starting from the original form of the building, has modified the Reichstag into a modern parliamentary office building. Foster has added a huge glass dome above the top floor that suits the building perfectly, and incorporates modern technology and ecology into the architecture. The materials of the dome are mainly metal and glass. For its transparency and through the inverted cones, sunlight is reflected into the main hall. In the daytime, the sunshades are automatically positioned to

改造前的德国国会大厦
The Reichstag Before Reform

改造后的德国国会大厦
The Reichstag After Reform

let the sunlight in, while in the nighttime, they alter to reflect the indoor light outwards, which makes the dome into a lighthouse that brings about unique scenery to the nights of Berlin.

新穹顶的弧度与被毁的原始穹顶保持了高度的一致性，设计师运用了最新的材料和技术实现对原始建筑风貌的重新再现。除此以外，所有建筑物外墙体都完整地保留了下来，使得建筑基本保留了历史风貌，尊重了人们对这个城市宝贵的记忆。德国国会大厦的改造原则是在尊重和保存建筑原有肌理的基础上，运用最新的建筑和工程科技，使其成为标志性建筑，是一次具有重要意义的尝试。

The new dome remains highly identical to the one that has been destroyed. The designer applied the most cutting-edge materials and technology and preserved all exterior walls for preserving the building's original style and respecting people' s valuable memories to the city. The principle of the reconstruction is to modify the building, in the basis of respecting and preserving the original texture, into a landmark by using the latest technology of construction and engineering. The attempt was a meaningful one.

建于 1898 年的巴塞罗那斗牛场在禁止斗牛后，由理查德·罗杰斯设计改造成了一处大型商业性的娱乐休闲综合体。改造设计最引人注目的是悬浮于原建筑立面上的金属飞碟状“空中广场”，这个空间容纳了景观平台以及餐厅等服务设施，结构由可循环使用的木制梁柱构成，外表用镀锌金属板面。在建筑的底层与城市街道交汇处，理查德·罗杰斯还设计了一系列小广场与地铁站和邻近的胡安·米罗公园形成了方便的联系。从底层四个方向都可通过十字形的交通系统进入新建筑的中庭以及各楼层和空中广场。

After the Las Arenas Bullring built in 1898 was prohibited from bull-fight, Richard Rogers designed for its reconstruction into a large commercial leisure and entertainment entity. The most eye-

德国国会大厦穹顶
The Dome of The Reichstag

改造后的巴塞罗那斗牛场
Las Arenas Bullring Before Reform

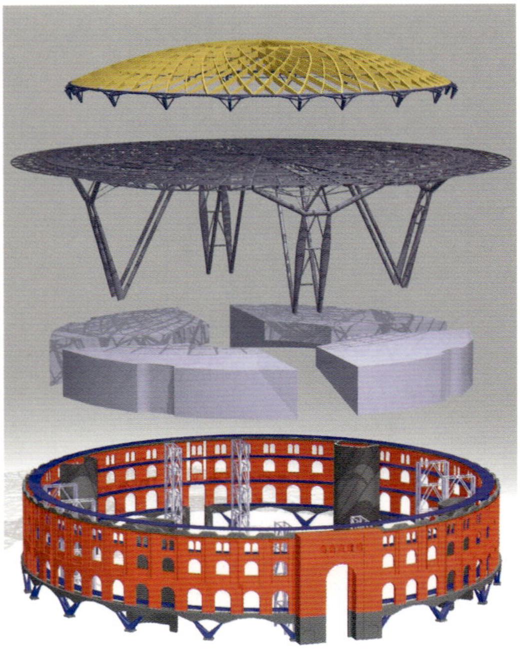

巴塞罗那斗牛场改造示意图
Reconstruction of the Las Arenas Bullring

catching part is the floating saucer-shaped circus built on the top that contains observation platforms, and restaurants. Structured by reusable wood beams and columns, its surface was covered by metal boards coated with zinc. Richard has also designed a series of small squares that connect with the metro station and the Joan Miró Park for public convenience. One can enter each level and the floating circus from entrance facing 4 directions and by the new building's cruciform transportation system.

斗牛场加建的顶棚形态灵感来自于公牛犄角明显上扬的曲线线条，显示了设计师对斗牛文化的尊重。一方面，斗牛场的内部虽然被完全拆除，改造成现代化的商业中心，但整个斗牛场的外墙体都保留下来，保存和修复了原有建筑的阿拉伯式立面，这使得建筑物的外立面最大限度地保留了原始风貌。另一方面，穹顶使用了高性能的玻璃进行日光照明，同时还结合被动式通风和太阳能热水装置，提供室内热水。因此改造后的斗牛场不仅充分展现了工业高科技之美，还在保留历史遗迹的前提下，将巴塞罗那新的一面和活力展现出来。

It was the upward rising curve of a bull's horn that inspired the designer to show his respect to the bull-fighting culture by adding the circular ceiling. On one side, though all internal were brought down and transformed into a modern commercial center, the entire outer wall has been preserved, and repaired including those arabesque vertical faces, thus preserving its original style as much as possible. On the other side, the dome was installed with high performance glass for daytime lighting, and passive ventilation system and solar water heater are also installed. Therefore, the reconstruction of the Bullring not only shows the beauty of industry and high technology, but also conveys the new energy of Barcelona while preserving its historical remains.

3 工业风格的当代发展
The Contemporary Development of Industrial Style

从 18 世纪 60 年代的工业革命开始，工业化创造了人类前所未有的文明，为了提高人类生活水平、加速文明进步、满足工业需求，而不断兴建更多、更大的工业厂房。随着时光流逝，科学技术的飞速发展，产业的更新换代使得曾经代表先进生产力的工业厂房逐步成为工业遗产。

欧洲的工业遗产保护运动保留了大量工业遗产，形成了“工业遗址地”和“工业遗址厂廊”，也成了当今工业文化旅游的必经之处。位于德国中西部的鲁尔区，形成于 19 世纪中叶，是典型的传统工业地域，被称为“德国工业的心脏”，是德国也是世界最重要的工业区。

From the Industrial Revolution of the 1760s, industrialization has created unprecedented civilization. In order to improve the standard of living of mankind, accelerate the progress of civilization, more and larger industrial plants were gradually built to meet the demand for industry. With the passage of time, and the rapid development of science and technology, the constantly upgrading industry has made the industrial buildings that once represented advance productivity power slowly became industrial heritage.

The industrial heritage conservation movement in Europe has retained a large number of industrial heritage which has remained the "Industrial Sites" and "Industrial Sites Gallery" that has become stops for every tourist of today's industrial culture tourism. Formed in the middle of the 19th century, located in the midwestern part of Germany, the Ruhr area is a typical traditional industrial area known as the "Heart of German Industry", and is also the world's most important industrial area.

↑ 鲁尔博物馆
Ruhrlandmuseum

埃森市是鲁尔区工业城市群中人口超过 50 万的大城市，埃森关税同盟煤矿工业区是由建筑师弗里茨·舒普和马丁·克莱默于 1927 年至 1932 年为这里设计的工业建筑群，建筑师们根据对称性和几何学原理，在钢结构中大胆采用红砖外墙，这种建筑式样时至今日仍然符合人们的审美需求。如今这里已由污染严重的老工业区改造成为埃森市的文化和艺术中心，同时也是展现采矿历史和工业建筑发展历程的一座鲜活的博物馆。参观者可通过保留下来的工业设施进入地下矿井，体验曾经的采矿环境，或搭乘昔日运煤火车游览并参与各种活动，实现了工业建筑的活力再生。

Essen is a large city with a population of more than 500,000 in the area. The Zollverein Coal Mine Industrial Complex in Essen is an industrial building group designed by the architects Fritz Schupp and Martin Kremmer from 1927 to 1932. They boldly used red brick walls with steel structures based on symmetry and geometrical principles. Today, architectural styles as so still meet people's aesthetic needs. Today it has been transformed from a heavily polluted old industrial area into the cultural and artistic center of the city of Essen, as well as a vibrant museum that shows the history of mining and the development of industrial buildings. Visitors can enter the underground mine through the retained industrial facilities, experience the mining environment, or take the coal train tour and participate in various activities. These has revived the industrial buildings.

鲁尔博物馆建于 100 年前，原为埃森关税同盟煤矿工业区中的洗煤塔，后由荷兰建筑师雷姆·库哈斯主持设计并改造，通过对原有矿业厂房建筑空间的组合与重构，展示了鲁尔区的历史与转型发展。当年运煤的传送带被改造成 55 米长，24 米高的自动观光扶梯，将游客带入到博物馆游客中心。扶梯采用富有现代感的黑色和橙色组合，一方面展示出新增设施与历史厂房的鲜明对比，另一方面蕴含着特殊的寓意，黑色代表煤矿，而橙色象征着熔炼时产生的红色灰烬。

鲁尔博物馆入口扶梯
Entrance Escalator of Ruhrlandmuseum

鲁尔博物馆的楼梯 © Thomas Willemsen Stiftung Zollverein
Stairs of Ruhrlandmuseum

Founded 100 years ago, the Ruhrlandmuseum was originally a coal waching tower in the area. It was later redesigned by the Dutch architect Rem Koolhass through combining and restructuring the space in the original mining buildings, which shows the history and transformation of the area. The conveyor belt of the old days was modified as a 55 meters long, 24 meters high automatic sightseeing escalator that carries visitors to the museum's visitor center. The escalator uses a modern color combination of black and orange. It shows the sharp contrast between the new facilities and historical factory buildings, while it contains a special meaning: black stands for coal, and orange symbolizes the the red ashes produced during melting.

游客中心以下的楼层展示了鲁尔区的自然、人文和改造历程。展示空间和展示内容的设计尊重和延续原有空间格局，内部大部分机械、混凝土和钢构件得以保留和展示，声音、图像、影像、实物、互动体验等展示方式综合运用，充分激发参观者的好奇心、探索欲和热情。

The floors below the visitor center show the nature, culture and transformation of the Ruhr area. Most of the internal machinery,

鲁尔博物馆内部
The Interior Space of Ruhrlandmuseum

concrete and steel components are retained and displayed. The design of the displaying space and the content respects and continues the original space layout. The integrated application of sound, images, videos, real objects, and interactive experience fully stimulates the visitors' curiosity, exploration and passion.

矿区原锅炉房由英国著名建筑师诺曼·福斯特主持并设计，改造成了如今在设计界享有盛名的红点设计博物馆。设计师保留了原有的高大的锅炉设备钢架结构，斑驳的砖墙、生锈的铁质构件，粗犷而强悍。与改造后穿插其间的现代工业材料、玻璃、抛光的合成材料、合金钢具、灯具等相结合，使历史与现代、工业与时尚、建筑与景观形成呼应。进入二层大厅，在足有三层楼高的空间中，悬挂着一辆现代工业的标志物——奥迪 A8 全金属车体，它的两侧一边是废弃的高大锅炉设备，一边是砖墙高筑，其中展示着历年获得设计界“奥斯卡”——红点设计大奖的当代设计产品。从餐具到灯具、从电子产品到厨卫设施、从服装服饰到交通工具，博物馆里陈列了大约 1000 件来自世界各地、各式各样的产品。

红点设计博物馆 © Red Dot Design Museum
Red Dot Design Museum

红点设计博物馆斯托克站的奥迪 A8 轿车
© Red Dot Design Museum
Audi A8 in the Stoker's Station of the Red Dot Design Museum

红点设计博物馆中的 D2V2 灯光设备 © Red Dot Design Museum
"D2V2" Light Installation in the Red Dot Design Museum

The original boiler room was designed by the famous British architect Norman Foster, and was transformed into the Red Dot Design Museum now renowned in the design community. The designer retained the steel structure of the large original boiler equipment, mottled brick walls, and the rusty iron components, rough and doughty. In contrast, the delicacy and exquisiteness shown by the intervened modern industrial materials, glass, polished synthetic materials, alloy steel, lamps, etc. have generated echoes between history and modern, industrial and fashion, and architecture and landscape. Into the hall of the second floor, in this three-storey space, hangs a modern industrial signs - Audi A8's all-metal body, on one sides of which sits large abandoned boilers, on the other side stands a high brick wall, on which the contemporary design are displayed - those have won the Red Dot Design Award （as known as the "Oscar" in design industry） over the years. From tableware to lighting, from electronics to kitchen and toilet facilities, and from clothing and accessory to transportation, the museum displays about 1,000 products from around the world.

红点设计博物馆中锅炉顶 © Red Dot Design Museum
Boiler Roof in the Red Dot Design Museum

红点设计博物馆大厅 © Red Dot Design Museum
Ground Hall in the Red Dot Design Museum

北杜伊斯堡位于鲁尔区西部的下莱茵，是鲁尔区的重要工业城市，也是欧洲最大的钢铁中心。北杜伊斯堡景观公园被认为是后工业景观公园的代表作，它也是工业建筑改造历史上跨时代的飞跃。从此以后，把工业建筑原汁原味保护下来，并激活重生的方式日超流行。景观设计师彼得·拉茨保留了工厂中所有的建筑以及工业配件，赋予部分构筑物新的使用功能，最大限度地保留了工厂的历史信息。并且利用原有的“废料”塑造公园的景观，减少对新材料的需求，保护工厂中的植被，从而建构出一幅工业文明与自然社会和谐共生的美好场面。

Located in Bas-Rhin in the west, North Duisburg is an important industrial city in the Ruhr area and the largest steel center in Europe. The Landschaftspark Duisburg-Nord is considered a representative work of the post-industrial landscape park. It is also a leap in the history of industrial buildings. Since then, the authentic form of the industrial architecture has been preserved, activated, and popularized. Landscape designer Peter Latz retains all the buildings and industrial parts in the factory, giving new features to some of the structures and maximizing the history of the plant. He also used the original "waste" to shape the park landscape, reduce the demand for new materials, and protect the vegetation in the factory, thereby constructin a beautiful scenery of industrial civilization and natural society in harmony.

↑ 北杜伊斯堡景观公园 ©kaвstn DiskCat
North Duisburg Landscape Park © kaвstn DiskCat

工业风格的设计理念十分适合于阁楼式公寓的设计，“LOFT”原指阁楼、顶层楼，也指工厂或仓库的上部楼层。阁楼式公寓起源于早期的低收入住房改造，首先出现在 20 世纪 40 年代的纽约。当时的艺术家们为了逃避市区高昂的租金，开始利用曼哈顿西南区废弃的工业厂房，从中分隔出居住、工作等不同的生活空间。厂房内开敞的空间、高耸的顶棚和宽大的窗户满足了艺术家艺术创作的需求，为他们自由建构各种生活形态提供了可能。

The idea of the industrial style is ideal for the Loft Style Apartment. The word LOFT refers originally to attics, top floor, and the upper floor of a factory or warehouse. The attic apartment was originated in the early transformation of low-income housing, first appeared in the 1940s in New York. At that time, artists began to use the abandoned industrial factory buildings in Southwest Manhattan in order to escape from the high rent in the city, by separating the buildings into residential, working and other living areas. With high roofs and large windows, these buildings offer openness to the artists that meets the needs of their artistic creation, and also provide the freedom for them to many possible life styles.

苏荷公寓
Soho Lofts

纽约的苏荷区在城市发展与文化保留的对立之间被保存下来，它的复兴为许多处于转型期的现代化城市提供了一个成功的范本——如何在工业遗产上进行文化创意产业园区建设。设计师采用“以旧做旧”的改造方案，在不破坏建筑立面造型的前提下，对工业建筑内部空间设施进行改造，用作艺术创作、展示交流的场所，同时辅之以餐饮业、旅游业等，使之成为艺术家公寓和文化创意产业的聚集区，形成了以艺术经营为主，同时融入其他产业的开发模式。

New York's Soho district was preserved despite the conflict between urban development and cultural preservation, and its

纽约苏荷区街道景观
New York Soho Street View

revival provided a successful model for many modern cities in transition about how to develop cultural and creative industry zones on heritage industrial sites. Designers followed reconstruction principle of "to create based on the old", and transformed the interior space of the buildings under the premise of not damaging any facade modeling. The buildings were remade for the purpose of artistic creation, display, and communication, and they have been supplemented by catering, and tourism and so on, making it a gathering place of the Loft artists and cultural and creative industries, and forming a development model that focuses on art business, while integrated with into other industries.

工业风格代表着这个时代留下的痕迹，它满足了当代人生活方式需求，采用现有的物质技术手段，体现和这个时代精神相和谐的价值观和审美观。所以工业风格应用于室内设计时都注重环境空间中地域文脉的表达，在满足空间已有的功能需求上，创造与人文、自然景观环境相融合的空间形式，从而体现工业风格独特的美丽。

Industrial style represents the traces of this era. It satisfied the needs of contemporary lifestyle, utilized existing material and technical means, embodied the values and aesthetics that coincides with the spirit of this era. Therefore, the industrial style, when applied to interior design, always pay close attention to the expression of the local culture among certain environment or space. Under the precondition of meeting the functional requirements for the existing space, a spatial forms that integrate with the landscape of culture and nature are created to reflect the unique beauty of the industrial style.

20 世纪末，中国开始出现改造和利用工业建筑为文化空间的现象，上海、苏州河畔大批的艺术家工作室、昆明的创库、四川美术学院的坦克仓库、北京的艺术加油站和 798 艺术区等。这些改造后的工业建筑，外表依然沧桑，充满了厚重的历史感，但是室内却是一个截然不同的艺术空间：黑色的楼梯、纵横的钢索、错落的空间结构、简洁明快的线条。工业风格要求尽可能保留原有工业厂房的建筑元素，保

北京 798 艺术区
Beijing 798 Art District

北京 798 艺术区
Beijing 798 Art District

留原来那些错落、斑驳的砖墙、纵横的管道，甚至保留墙壁上那些具有年代特色的朱红标语以及部分工业机械部件。空间内当代的艺术作品、舒适的现代家具，与斑驳的墙面、过时的机械等历史痕迹相映成趣，这些旧空间经过改造后已经成为了新的建筑作品，它们在历史文脉与经济发展中、在实用与审美之间与旧建筑展开了生动的对话。

At the end of the twentieth century, China began to utilize and transform industrial buildings into cultural space. This phenomenon was presented by a large number of artists workshops near the river bank in Suzhou, Shanghai, the Creativity Library in Kunming, the Tank Warehouse in Sichuan Academy of Fine Arts, and the Art Station and 798 Art District in Beijing. After the modification, these industrial buildings still appear to be vicissitudes and full of heavy sense of history, but the interior is a very different art space: Black staircase, vertical and horizontal cable, scattered space structure, simple and neat lines. The industrial style requires the architectural elements of the original industrial buildings to be kept as much as possible, retaining the original scattered, mottled brick walls, vertical and horizontal pipelines, and even the old wall-painted slogans and some industrial machinery parts. Inside, the contemporary art works, comfortable modern furniture, mottled walls and outdated machinery and other historical traces formed a delightful contrast. These old space has been transformed into new architectural works, which have launched a lively dialogue between the old buildings in the context of historical culture and economic development, and practicality and aesthetics.

北京 798 艺术区内部墙面保留的标语
The Reserved Slogan on Interior Wall of Beijing 798 Art District

第2章

Chapter 2

工业风格的装饰元素

The Decoration Elements of Industrial Style

1 天花、墙面与地板

Ceiling, Wall and Floorboard

天花 Ceiling

工业风格起源于旧厂房、仓库的改造，因此工业风格的天花基本不做任何修饰，尽量保留原来的结构，包括梁、柱和管道。最简单直接的方式就是保留不加装饰的水泥顶，如果担心掉粉尘颗粒，可以刮平做细致一些，然后直接刷白、喷黑、喷灰或者刷任意喜欢的颜色，裸露的管道也可以喷成喜欢的颜色。需要注意的是，虽然裸露的天花不会降低层高，且造价较低，但是因为没有吊顶的遮盖，管线直接裸露在外，所以应提前对管线进行规划布局，以免过于凌乱，影响美观，如果有需要也可以用旧木料做天花或者假梁。

Industrial style is originated in the old factory, the transformation of the warehouse, so the ceilings of industrial style are usually remained undecorated, due to the attempt of retaining the original structure, including beams, columns and pipes. The simplest and most straightforward way is to keep the cement ceiling without decoration. If worried about dust droping, fine scraping can be done in order to avoid it, and then the ceiling may be painted or sprayed in white, black, gray or any other favorite colors, the exposed pipes can also be sprayed in the favourite colors. However, it should be noticed that although the bare ceiling will not reduce its height, and is of lower cost, its status of being unconvered and the exposure of pipelines demand planning for layout in advance, so as not to be too messy in appearance. Old wood can be used to build plafond or false beams if necessary.

原木假梁天花。摄影：Edgardo Contreras
Log False Beems, Photographer: Edgardo Contreras

灰色毛坯水泥屋顶、裸露的金属水管、局部木饰面天花造型和简洁的光管，营造出一个充满科技感的空间
The Gray Rough Concrete Roof, Bare Metal Pipes, Local Wood Ornamented Ceiling Structure and Simple Light Pipes Created a SpaceFull of Science and Technology

墙面 Wall

墙面一直以来都是空间装饰的重点，对风格的塑造起着至关重要的作用。刻意保留未经装饰的墙面是工业风格的重要特征之一，保留裸露的砖墙或者毛坯水泥墙面，有时候甚至需要人为的对墙面进行一些破坏，以达到做旧磨损的效果。砖墙可以粉刷成白色或者黑色，但一定要保持其原始的肌理。比起砖墙的复古感，水泥墙更有一分沉静与现代感，不一定非要刷的均匀平整，手工造成的随机纹理也有不错的效果。工业风格墙面以有粗糙肌理的材料为主，常用的墙面装饰材料有：文化石、红砖、水泥、水泥板、马赛克拼贴、仿古墙面漆、粗糙的木板等。

The wall has always been the focus of decoration, which plays a vital role for styling. Deliberately reserving undecorated walls are one of the important features of the industrial style. Bare brick walls or rough concrete walls are retained, and sometimes it is required to deliberately damge the surface in order to achieve worn and distressed effects. The brick wall can be painted white or black, but its original texture must be kept. Compared to the retro sense of the brick wall, the cement wall is more calm and modern. It does not have to be make very even or smooth, since hand-made random texture has quite good effects. Materials of the industrial style walls are basically rough in texture, and these are some materials commonly used: cultural stones, red bricks, cement, cement boards, mosaic, antique wall paint, rough wood boards and so on.

两面墙选择了不同的材质，因为色调的统一，所以并不突兀。摄影：Matteo Favi Immagini

Different Materials was Chosen for the Two Walls, and Due to the Identity of Color Tone, It Does not Look Abrupt, Photographer: Matteo Favi Immagini

保留原建筑的红砖墙，白色墙裙让我们回忆起 80 年风靡一时的绿色墙裙，充满复古气质。摄影：Dave Wheeler

The Retained Red Brick Wall of the Original Building, the Old-fashioned Skirted White Wall Brings Back Memories of the 80s, When the Green Wall Skirt was Popular at the Time, Photographer: Dave Wheeler

地板 Floorboard

地板除了使用功能以外，对室内氛围和风格的塑造也起着至关重要的作用。在工业风格空间中常用的地面材料有：仿古砖、仿古地坪漆、仿古木地板等。仿古砖是常见的地面铺装材料之一，但是并不是所有的仿古砖都适合，在挑选仿古砖时要挑选色彩纯度较低的有做旧肌理的样式，例如模仿铁锈肌理的铁锈砖；仿古地坪漆是一种特殊地坪涂料，能与水泥起作用形成一种做旧的肌理效果；木地板要选择颜色比较灰暗，有明显的肌理和做旧效果的仿古木地板。

In addition to its functions, the floorboard also plays a vital role to the interior atmosphere and style shaping. These are some common materials used for the floors in the industrial style space: Antique tiles, antique floorboard paint, antique wooden floorboard and etc. The antique tile is one of the common ground pavement materials, but not all antique tiles are suitable. In the selection of antique tile, it is necessary to choose old textures in low-purity colors, such as the rust texture resembling bricks; Antique floorboard paint is a special floorboard paint, which can create old texture by relating with the cement. And for the wood floorboard, it would be good to choose boards in relatively dark colors, with obvious texture, and distressed.

仿古砖
Antique Tiles

仿古木地板
Antique Wood Floorboard

水泥地面刷仿古地坪漆。设计：Chanou
Cement Floorboard Brushed with Antique Floorboard Paint, Designed by Chanou

Z 色彩
color

营造工业风格的关键之一就是色彩的搭配，工业风格设计本身体现着颓废、硬朗、个性的感觉，这种风格的色彩塑造以中性色为主。色彩与材料本身的质感也有很大的关系，相同的颜色在不同质地的材料上所呈现出来的效果是不一样的，如：光洁的金属材料刷漆表现出的是鲜亮的色彩，砖墙、水泥墙刷漆表现出的则是质朴的色彩。因此在进行色彩设计时应充分考虑材质本身的特性对色彩表现产生的影响。

One of the keys to create an industrial style is color matching. The industrial style design itself reflects feelings of decadency, toughness, and personalization, and the colors of this style are basically neutral. The colors and the texture of the material itself have much related, and even the same colors on different textures present non-identical effects, which can be exemplified as: Brush painting on bright and clean metallic materials shows the brightness of colors, while on brick walls and cement walls, it presents simplicity. Hence it should be fully considered, during color design, that the natures of the material itself will impact the presentation of colors.

这是一个办公空间的员工就餐区，设计师以大面积的白色为主色调，搭配黑色的金属工业吊灯和座椅，简洁理性，特别适合办公空间。摄影：Ilya Ivanov
This is A Staff Dining Area of An Office Space, Where The Designers Used a Large Area of White, Along with Black Metal Ceiling Lamps and Seats. The Style is Simple and Rational, Especially Good for Office Space, Phottgrapher: Ilya Ivanov

黑、白、灰经典搭配。黑色沉稳，白色优雅，搭配不同深浅的灰，可以创造出更多层次的变化。选用纯粹的黑、白、灰色系，更容易体现出工业风格硬朗的特质。工业风格墙面常常使用大面积的冷色调来营造气氛，用砖墙取代经过粉饰的光滑墙面更具有复古的韵味，砖块与砖块之间的缝隙呈现出有别于一般墙面的光影层次效果，将砖块粉刷成黑色、白色或者灰色，可以为室内带来既复古又摩登的视觉效果。

The classical matching of black, white and gray. the black shows calmness, white shows elegance, and with different shades of gray, you can create more levels of change. It also makes it easier to reflect the characteristics of the industrial toughness. Industrial style walls often use large areas of cool colors to create different atmospheres, and using brick walls instead of smooth walls creats more retro charm. The gaps among the bricks show the light and shadow effects that are different from ordinary walls. To brush paint the bricks into black, white or gray, can bring retro, while modern visual effects to the interior space.

砖红色、灰色、原木色等自然色调。工业风格强调材质的本质，如红砖墙的红色、素水泥的灰色、金属色，也可能是铁锈色。比如红砖墙，给人质朴、温暖、欢快之感，但是单纯的大面积的使用红砖墙也会感觉色彩过于单一，让人产生沉闷等不适感，这时，可以用白色、灰色与砖红色相配，不仅可以打破沉闷，还可以更好的衬托红砖墙的独特魅力。

Brick red, gray, wood color and other natural colors. The industrial style emphasizes the nature of the materials, such as the red of a brick wall, or the gray, metallic, and rusty color of plain cement. For instance, a red brick wall gives a rustic, warm, and cheerful feeling, but to simply use them in a large area would make it overly dull, and bring boring and discomforting effects. In this case, a combination of white, gray and red bricks will not only break the dullness, but also enhance the unique charm of red brick wall.

大胆使用高饱和点缀色。工业风格的室内设计由于其采用的元素基本以冷色调为主，长时间处在其中，会给使用者一种冰冷的视觉感受。因此，可以使用一些饱和度高的颜色作为点缀，例如红色和黄色。

Bold use of high saturation embellishment colors. Industrial style interior design, to a certain extent, will give cold visual experience to a long time user, because of its primary applying cool colors on the elements. Therefore, some high saturation colors can be used as embellishment, such as red and yellow.

原木的温润质感，让这个冷漠、机械的空间多了一抹温暖的色彩，墙上挂的带有生活气息的复古黑白装饰画，则为这个空间注入了一份活力。摄影：Ilya Ivanov
The Mild and Humid Texture of the Logs, Brings a Touch of Warm Colors to this Indifferent and Mechanical Space. Hanging on the Wall, the Retro Black and White Decorative Paintings Delivers a Breath of Life, and Adds Vitality to the Space, Photographer: Ilya Ivanov

红色是消防色的一种，也是工业风格空间最常用的点缀色。摄影：James French

Red is One of the Fire-fighting Style Colors, and Common in Industrial Style Space as Decoration, Photographer: James French

增加不同的点缀色提亮，空间会呈现出不同的格调韵味，如使用绿色点缀，就能赋予室内生机与活力。摄影：Marcin Ratajczak

Joining With Different Embellishment Colors to Intensify Brightness, the Space Will Flourish a Different Style of Charm. If Rendered With Green Embellishment, Vitality and Dynamism Will be Given to the Room, Photographer: Marcin Ratajczak

为空间增加温暖色彩的暖光源。暖光源的照明可以中和工业风格单调、冷硬的氛围。无论具有温柔黄光的白炽灯还是色彩斑斓的霓虹灯，都是点亮室内空间的最佳选择。

Integrating warm light sources to the space. Warm light can neutralize the uninteresting, cold and tough atmosphere of the industrial style. Whether it is gentle yellow lights or vibrant neon lights, they are the perfect choices for lighting the interior space.

暖色调的原木色和灯光中和了工业风格的冷硬氛围。摄影：Julian Ward

The warm shades of wood and light neutralize the cold and hard atmosphere of the industrial style, Photographer: Julian Ward

3 家具

Furniture

金属制成的家具无疑是工业风格中最富有特色的代表之一。金属不但是持久耐用的材料，更是现代工业化的标志。

Metal-made furniture is undoubtedly one of the most distinctive representatives of industrial style. Metal is not only a durable material, it is a sign of modern industrialization.

由迈克尔·索奈特于 1859 年运用蒸汽弯曲木技术制造而成的 14 号椅（No.14 chair）又称维也纳咖啡椅（Vienna Cafe Chair），是工业风格空间常用家具。这种圆背木椅没有扶手，造型修长，由一整根圆木料弯曲成一个弧形，一直从椅背连到后部的两条腿上，后期改为金属管制造。这把椅子被公认为第一批批量生产的椅子，更是 19 世纪最佳的家具设计之一。

Made by Michael Thonet in 1859 with the adoption of steam bending wood technology, the No.14 chair, also known as the Vienna Cafe chair, is a kind of furniture commonly used industrial style spaces. This round slender wooden chair has no armrest, the curve of a whole bent wood constructs the backrest and all the way down to the back legs. The chair was later manufactured with metal tubes. It was recognized as the first batch of mass produced chair, and it is also one of the best furniture design of the 19th century.

14 号椅
No.14 Chair

由托莱多金属家具公司（Toledo Metal Furniture Co.）于 20 世纪 30 年代设计制造并申请专利的系列工业绘图凳（Drafting Stool）是工业美学家具的典范。

The series of industrial Drafting Stool, designed and manufactured by Toledo Metal Furniture Co. in the 1930s, is an example of furniture of industrial aesthetics .

20 世纪 30 年代，托莱多公司生产的可调节高度的工业绘图凳
In the 1930s, Toledo Produced an Height-adjustable Drafting Stool

托莱多工业凳，表面漆带有裂纹
Toledo Industrial Stool , withAge Cracked Red Surface Paint

带有坚硬可调节金属基座以及弯曲木座椅和靠背的当代复古托莱多凳
Vintage-modern Toledo Stool Featuring Sturdy and Adjustable Metal Base with Bentwood Seat and Back.

扎维·博洽德是法国电镀设计先锋。金属构件经常锈蚀，因此扎维·博洽德设法用镀锌方法来保护金属产品，延长金属的氧化过程。他在 1927 年注册了商标 TOLIX，TOLIX 镀锌家具是典型的工业风格家具。TOLIX 镀锌家具中最著名的是扎维·博洽德于 1934 年设计的玛莱椅（Marais A Chair），玛莱椅采用铁皮做支架和椅面、靠背，再镀锌防锈，色彩多是灰色的。因为这把椅子比较笨重，1956 年扎维·博洽德的儿子让·博洽德对这把椅子的设计进行了修改，椅子变得更轻便、雅致、修长。让·博洽德将这把新椅子命名为 A56 号椅，这把椅子结实、舒适度高、可叠放、色彩也更加丰富，特别适合法国的小酒馆、小咖啡馆使用。

Xavier Pauchard is the pioneer of the French electroplating design. Metal components are often rusted, so Xavier Pauchard seeks to use galvanized methods to protect metal products and extend the oxidation process. He registered the trademark TOLIX in 1927, the galvanized furniture of the brand is a typical industrial style furniture. Of all the TOLIX galvanized furniture, the Marais A Chair designed in 1934 by Xavier is the most famous. It uses iron sheet to make the frame, surface and backrest, and then galvanized, so the color is mostly Grey. Since the chair was relatively heavy, Xavier Pauchard's son Jean Pauchard, modified the design of the chair in 1956 to make it lighter, more elegant and slenderer. The new one was named as A56 Chair by Jean. Its toughness, comfort, stackableness, and variation of color make it especially suitable for French bistros and small cafes.

玛莱椅
Tolix® Marais A Chair

A56 号玛莱扶手椅
Tolix® Marais A56 Armchair

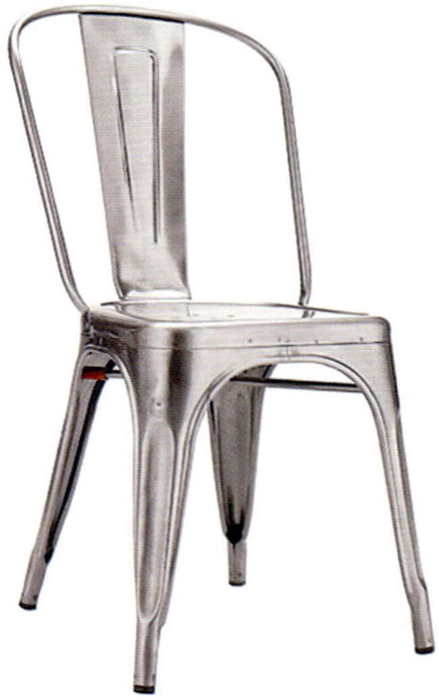

玛莱吧台凳
Tolix® Marais Barstool

简·普鲁威是一位擅长以金属为主体材料制作家具的现代设计大师。从 1924 年起他开始充分利用刚发明不久的电焊技术制作金属薄板家具，其设计中强烈的现代工业美学气息立刻吸引了一批前卫设计大师的注意， 其中包括勒·柯布西耶及其事务所的其他设计师。普鲁威的设计极其新颖而大胆，他时常结合机械装置设计出各种可调节的椅子来，并多次获得奖项。

Jane Prouve is a modern design master of metal-maken furniture. From 1924 onwards he began to make full use of the recently invented welding technology to produce metal sheet furniture. The strong modern industrial aesthetic atmosphere in his design immediately attracted a number of avant-garde design masters, including le Corbusier and other designers in his firm. Prussian's design is extremely new and bold. From time to time, he designed a variety of adjustable chairs with the idea of machinery and won several awards.

↑ 简·普鲁威于 1930 年设计的标准椅
Standard Chair, Designed by Jean Prouve in 1930

↑ 简·普鲁威于 1931 年设计的安东尼椅
Antony Chair,Designed by Jean Prouve in 1931

↑ 简·普鲁威于 1950 年设计的 EM 桌，特别适合作为餐桌或工作台使用
EM Table, Designed by Jean Prouve at 1950, Suitable as Dinner Table or Work Table

由埃美柯（ Emeco ）于 1944 年生产的海军椅（Navy Chair）也是典型的工业风格座椅。埃美柯是美国的一个家具公司的品牌，主要生产座椅。

The Navy Chair produced by Emeco in 1944 is also a typical industrial styled seat. Emeco is a brand of a furniture companie in the United States, specializing in the production of seats.

海军扶手椅
AVY® ARMCHAIR

海军吧台凳
NAVY® BARSTOOL

海军椅
NAVY® CHAIR

奥门克斯顿卡椅（Omkstak chair）是英国设计师罗德尼·金斯曼于1971年设计的，是20世纪70年代最为流行的椅子之一。这把椅子的特色在于，它在整条钢管弯曲而成的框架内安装上压制钢板的座板和靠背，并在上面镂刻出整齐的圆孔。这种椅子被漆成各种色彩，不仅拥有经久的审美价值，而且功能性极强，可用于室内或者室外，可以叠堆起来，也适于成排置放。

Omkstak chair is designed by British designer Rodney Kinsman in 1971 and is one of the most popular chairs in the 1970s. The chair is characterized by the fact that pressed steel plates with neatly engraved round holes are installed on the frame of a single bent steel pipe as the seat plate and backrest. This chair is painted in a variety of colors, not only having a long aesthetic value and highly functional, it can also be stacked up or placed in line which is fit for both indoors and outdoors.

奥门克斯顿卡椅
Omkstak Chair

金属是工业风格家具最常用的材质，但是金属冰冷的质感，如果大量用于生活中会显得过于冷硬。将金属与木材做混搭，既能增加生活气息又不失粗犷韵味，木材最好选择带有自然肌理的原木，甚至是回收的老木料，这些老木料无论从木头本身的纹路还是经过时间沉淀的颜色来看，都给人一种历经沧桑的感觉，让我们可以从中窥探到大工业时代的记忆。

Metal is the most commonly used material for industrial styled furniture, but the metallic cold texture will bring too much coldness and hardness to our life if used a lot. Matching the metal and wood can increase the sense of vitality without losing the charm of roughness. the best choice of wood is the logs with natural texture, or even recovered old lumber, which, either the wood itself or the colors rendered by the passage of time, gives a feeling of vicissitudes, where we can look into the memories of the great industrial era.

实木与金属混搭的工业风格餐桌
Industrial Styled Table Mixed with Solid Wood and Metal

以工业时代为背景的工业风格空间，不可或缺的元素之一就是机械装饰构件。我们可以将这些装饰构件穿插到家具设计中，如常见的带轮子的茶几、边桌、转椅等。

In a space in the backround of the industrial age, one of the indispensable elements is the mechanical decorative components. We can put them into the furniture design, such as the common coffee tables with wheels, side tables, swivel chairs and so on.

用回收的旧木料与金属构件制作的可移动茶几
A Movable Coffee Table Made of Recycled Old Wood and Metal Components

略带温情的复古怀旧气息是工业风格设计最常用的主题。尽管工业风格空间中会有大量的金属元素，但并不强调金属的光泽，所以整体空间比较暗淡，且略显冰冷。经过表面做旧处理的深棕色或者黄棕色的皮革沙发的温润质感和恰到好处的纹理褶皱，恰恰中和了这种冰冷的感觉，且完美地诠释了时代的沧桑感，是塑造工业风格的最佳选择。

The warm retro and nostalgic atmosphere is the most commonly used theme among industrial style design. Although there are a lot of metal elements in the space, the metallic luster is not emphasized, so the overall space is bleak, and slightly cold. Having been distressed, the warm texture and the proper fold of a dark brown or a yellow brown sofa just neutrualized this coldness, and perfectly interpreted the vicissitudes of the times, and is the best choice for molding industrial style.

复古做旧真皮沙发
Retro Distressed Old Leather Sofa

4 灯具
Lights

电的发明和使用给人类的生产和生活带来巨大的变化，也带动了其他工业部门的迅速发展。英国工程师斯旺以及美国发明家爱迪生共同推进了白炽灯的发明和应用，1879 年爱迪生成功研制出碳化纤维（即碳丝）白炽灯，率先将电光源送入家庭。1880 年爱迪生又研制出碳化竹丝灯，使灯丝寿命大大提高，同年十月爱迪生在新泽西州自行设厂，开始进行批量生产，这是世界最早的商品化白炽灯。早期的白炽灯以暖黄色调为主，这也是工业风格空间多选用暖光光源的原因。

The invention utilization of electricity has brought great changes to human life and production, and has also led to rapid development of other industrial areas. British engineer Swan and American inventor Edison jointly promoted the invention of incandescent. In 1879, Edison successfully developed carbon fiber （ie carbon wire） incandescent, which brought the first electric light into family. In 1880, he developed carbonized bamboo light that greatly improved the filament life. He set up his own factories in New Jersey in October, and began the mass production of the world's first commercial incandescent. Early incandescent lamp was basically yellow or warm-colored, which has lead to the major choice of warm lighting in industrial style space.

复古钨丝灯泡，在细节与工艺的处理上均还原了当时的制造水平和审美，迷恋工业风格的人们一定对各式的钨丝灯泡情有独钟
The Retro Tungsten Filament Lights that Restores the Manufacturing Level and Aesthetic at the Time in Terms of Details and the Processing. Those Obsessed with the Industrial Style Will Surely be Fascinated by Various Kinds of Them

工业风格灯具的材质通常为金属材质，并带有铁锈或者磨损痕迹。早期的工业照明灯具都有一个用于保护灯泡的金属罩，因此金属灯罩成为工业风格灯具的典型特征之一。工业风扇与灯具结合而成的风扇灯也常应用于工业风格的室内空间中。用金属管或者钢筋自己动手制作灯具，可以制造出别具一格的灯具，既环保又有创意。

Industrial style lamps are usually made of metal and marked with rust or abrasion. The early industrial lighting fixtures have a metal cover for protecting the bulbs, so metal shades has become one of the typical features of the style. The fan lights, being a combination of industrial fans and lights are often used in industrial style interior space. Unique DIY lights can also be made by metal tubes or steel bars, both eco-friendly and creative.

灯泡与金属风扇罩组成的吊灯
Ceiling Lights Comprising Light Bulbs and a Fan Cover

灯泡与金属灯架组成的吊灯
Ceiling Lights Comprising Light Bulbs and a Metal Bracket

灯泡与水龙头组合成的壁灯
A Combination of Light Bulbs and Faucets

灯泡与水管组成的艺术台灯
An Artistic Lamp Made by a Light Bulb and Water Pipes

由法国工程师伯纳德·阿尔滨·格拉斯于 1922 年设计的机械手臂灯 La Lampe Gras，不用焊接也没有螺丝，将工程力学的精巧技术发挥到极致。原为办公或工业环境下使用，后来逐渐变成了一种在家居和商业空间中都越来越普及的摇臂灯。

Designed by the French engineer Bernard Albin Gras in 1922, the mechanical arm lights La Lampe Gras has no welding or screws. This has brought the sophisticated technology of the engineering mechanics to the limit. Arm lights - formerly used for office or industrial environment, and later became increasing popular for homes and commercial space.

机械手臂灯 La Lampe Gras
La Lampe Gras Mechanical Arm Lights

与 La Lampe Gras 风格相似的机械手臂灯 Jielde Lamp，由法国设计师让·路易斯·多梅克在 1948 年设计，是法国工业风格灯具代表之一。此款台灯用刹车盘做底座，铁管两端是关节半球，球状关节部位接触面使用黄铜环导电，可以折成任意角度和高度。

Similar to the La Lampe Gras style, the arm light - Jielde lamp, designed by French designer Jean Louis Domecq in 1948, is one representing the French industrial style lamps. This table lamp uses a brake disc as its base. At each end of the iron tube, there are two ball joints with conductive contact surface inside, so it can be adjusted into any angle and height.

机械手臂灯 Jielde Lamp
The Jielde Mechanical Arm Light

诞生于 1932 年的 Anglepoise 可以说是英国悬臂式台灯的先驱，是 20 世纪最经典的灯具之一，也是设计史上被复制、模仿最多的产品之一。设计师乔治·卡沃迪恩是一位汽车工程师，他在 1932 年偶然发明了一种新型弹簧，无论把它往那个方向拉，它都能恢复到原始的刚直状态，最终它出现在了台灯上。开始卡沃迪恩针对的是工厂工人的需要，后来发现 Anglepoise 悬臂式台灯一样适用于办公室。

Founded in 1932, the Anglepoise can be said to be the pioneer of the British cantilever lamp, one of the most classic lamps of the 20th century, and one of the most imitated products in the history of design. Its designer George Carwardine, an automotive engineer, accidentally invented a new spring in 1932 that whichever direction it is pulled, it can restore itself into the original straight state, and eventually it appeared on the lamp. At the beginning, Carwardine aimed at the needs of factory workers and later found that Anglepoise was the same suitable for offices.

Anglepoise 悬臂式台灯
Anglepoise Cantilever Table Lamp

以金属为主体材料的设计师普鲁威设计新颖大胆，其于 1950 年设计的 Potence 系列灯具，以极度简约的线条，将人们的视线引向灯光本来的面貌。

Jean Prouve, a bold, metal-based designer, designed the Potence series of lamps in 1950, with extremely simple lines that drew people's attention to the original appearance of light.

大厅悬臂灯
Potence Wall Light

比起 Potence，265 壁灯（265 Wall Lamp）除了一样可以平贴于墙面节省空间外，还可以根据照明的角度上下调整，适合于多功能空间以及工作室空间使用。

Compared to the Potence, the 265 Wall Lamp, which can be attached flatly on walls to save space, is also vertically adjustable for right lighting angles, hence suitable for multi-purpose spaces and studio spaces.

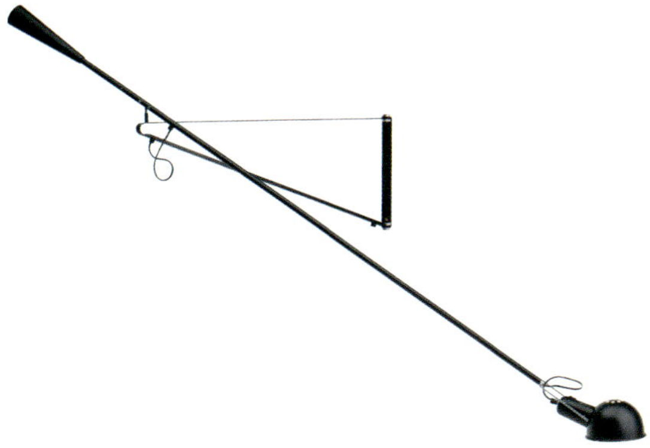

265 壁灯
265 Wall Lamp

英国 Original BTC 是 1990 年由彼得·鲍尔斯创立的。Original BTC 旗下还有两个品牌，Davey Lighting 和 LED lighting。Original BTC 的所有品牌都从英国丰富的后工业化历史中汲取灵感，结合最新的现代技术和复古技术来创造他们的设计。

The British original BTC was founded in 1990 by Peter Bowles. It has two sub-brands: Davey Lighting and LED lighting. All of them were inspired by the rich, post-industrial history of the UK, and combined with the latest modern and antique technology to create their designs.

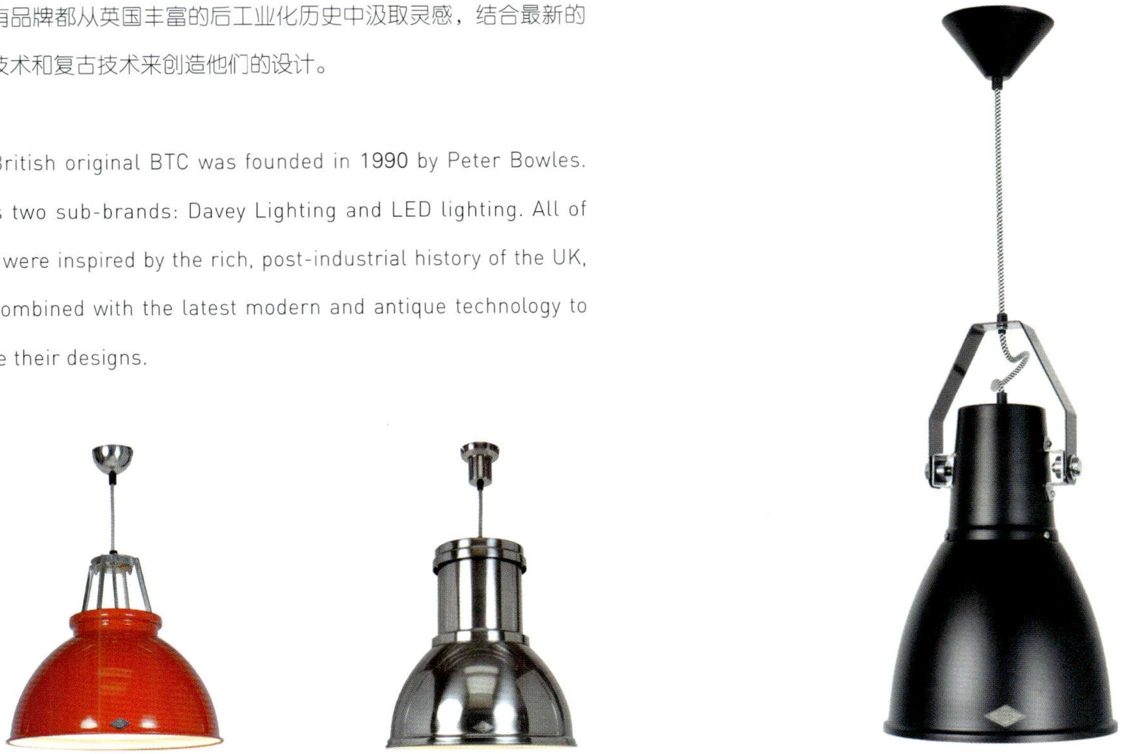

Original BTC 吊灯，适用于餐厅以及工作空间
Original BTC Chandelier Applies to Restaurant and Work Space

蜘蛛灯（Dear Ingo）也是在工业风格空间中经常出现的灯具，是设计师罗恩·吉拉德的作品，有黑白两个颜色。设计师将伸臂式台灯与吊灯充分的结合在了一起，看起来像一只悬挂的蜘蛛，你可以随意调节吊灯的高低，可以随意转动各个灯头的方向，让人觉得这盏吊灯也具有非凡的生命力。

The Dear Ingo is also a commonly seen type of lights in the industrial style spaces, which was designed by Ron Gilad, in the colors of black and white. The extended arm lamp and the chandelier are fully combined together, which makes it look like a hanging spider. You can freely adjust the chandelier level, or turn each lamp facing different directions. This makes people think that the light has extraordinary vitality.

蜘蛛灯
Dear Ingo

现代工业风格的灯具则与新材料以及新风格结合形成了独特的风貌。由英国设计师本杰明·贝尔设计的 Heavy Light 吊灯，让混凝土首次以灯具的身份优雅登场，打破了人们对混凝土的固有印象。此外，设计师设计的 Heavy Desk 台灯，还推出了了乳白的混凝土造型，洗刷了混凝土灰沉沉的印象。

Modern industrial styled lights have combined with new materials and styles to form a unique characterist. Heavy Light, designed by Benjamin Hubert, the British designer, has elegantly brought forth the concrete with the identity of luminaire for the first time, breaking the people's stereotypes of concrete. In addition, the designer of Heavy Desk lamp has also introduced a white concrete version that whitewashed the impression of the concrete gray.

Heavy Light 吊灯
Heavy Light

Heavy Desk 台灯
Heavy Desk

5 布艺
Fabric

工业风格所展现出来的机械、理性、冷硬的特点与纺织面料的感性、柔软相悖，因此许多布艺产品都无法与其相搭，例如带有帷幔的欧式窗帘、碎花图案的田园风格床品、抱枕等。如果想要将两种气质截然不同的物品搭配在一起，则需要找到两者之间的共性。

Among all the features of the Loft style, its hard, cold, rationalistic, and machinery-like characteristics just do not match with the sensibility and softness possessed by fabric, for instance, European curtains with valance, floral patterned country style bedding and cushions. To make the two kinds of items go together, it is essential to find out the similarities between them.

窗帘是居住空间必备品之一，它不仅能够保护个人隐私，还能有效地调节室内光线强度，减少室内的紫外线照射，营造舒适健康的室内环境。工业风格以蒸汽时代为蓝本，其间会加入许多机械元素来体现其复古风貌，一切与蒸汽时代有关联的事物都可以用在工业风格中，因此我们可以选择印有齿轮、地图、古老的报纸、蒸汽火车或双翼飞机等图案的窗帘布料；大量地运用金属元素也

设计师：Nima Keivani
Sina Keivani
Solmaz Fooladi
摄　影：Sergey Savchenko
Designer：Nima Keivani
Sina Keivani
Solmaz Fooladi
Photographer：Sergey Savchenko

是工业风格的一个特点，因此我们可以用一些有着相似光泽的窗帘面料，比如银色丝质质地的窗帘就是不错的选择；除了一些复古的元素，工业风格的空间中家具布置也比较开阔自由，因此我们也可以试着减弱窗帘的存在感，选择最普通的白纱帘，它对光线和视线都只起到了模糊的作用，浅淡自然的感觉既让人把注意力集中到它身上，又可以让室外的风景若隐若现，从视觉上扩大了空间感，同时这种色彩单一且透明的质地不容易影响到室内家具的搭配，是一种“打底色”一样的存在；百叶窗更具有一些现代气息，横向规则排列的肌理所呈现出来的质感与工业风格的气质相匹配，可自由调节的叶片角度能够使阳光在室内形成斑驳的倒影；简单淳朴的纯色窗帘则与工业风格质朴、注重功能性的一贯理念相匹配，颜色的选择需与空间的布置相互呼应。

设计公司：Josephine Hurley Architecture，摄影：Tom Ferguson
Design Studio: Josephine Hurley Architecture, Photographer: Tom Ferguson

Curtain is one of the necessities in a living space. It not only protects privacy of individuals, but also adjusts indoor lighting and reduces ultraviolet radiation, creating a healthy and comfortable interior environment. Industrial style models the age of steam by adding many machinery elements to express its retro characteristics. All related to that era can be adopted therefore we could choose the cloth on which pattern such as cog-wheels, maps, old newspapers, steam trains, or biplanes etc. The Loft style also features large appliance of metal textures. Thus we could choose glossy fabrics for curtains, such as one with silver and silky quality. Except for its retro elements, an industrial

设计公司：Egue y Seta，摄影师：Víctor Hugo
Design Studio: Egue y Seta, Photographer: Víctor Hugo

设计公司：Roy David Studio，摄影师：Yoav Gurin
Design Studio: Roy David Studio，Photographer: Yoav Gurin

featured space allows rather free and open furnishing, so we could also try to lessen the existence of the curtain by choosing white sheer curtains which merely blur the vision and light. The light and natural color draws one's attention to the curtain itself while indistinctly shows the view outside, thus broadens the spaciousness visually. At the same time, the single-colored and transparent texture acts as a background color that would not disturb the matching of other furniture. Because of its smell of modern times, a louvers match with the loft style well. Its regular, horizontally arranged shutters are adjustable, letting in sunlight and producing mottled shadows on the floor. The simple, unsophisticated single-colored curtain matches perfectly with the simple, yet practical ideas from the Loft style. Colors should be consistent with the arrangement of space.

工业风格空间中的床品只注重其实用性，没有传统的床裙或者床罩等。材质多为棉麻或者带有金属光泽的丝绸，局部点缀粗线编织的毛毯或者动物皮毛。色彩较单一，一般都是单色的床单、被子和枕头，色彩纯度较低，偶尔有较为鲜艳的色彩，质地也偏粗糙。

设计师：Ramon Zancanaro
Designer: Ramon Zancanaro

Loft style focuses only on utility, and does not have traditional bedding wares like bed skirt or bedspreads. The materials

设计师：Velizar Dimitrov
Designer: Velizar Dimitrov

摄影：Leonas Garbaciauskas
Photographer: Leonas Garbaciauskas

are generally cotton and linen, or silk with metallic gloss, interspersed by cable knit carpets or fur at some area. The colors are rather simple-usually there are bed sheets and pillows in low saturation, mono color, and made of rough material. Bright colors are rarely seen.

工业风格空间宜选择造型简洁的靠枕，正方形或者长方形是最佳选择。色彩多为纯度较低的中性色调，有时也用纯度较高的亮色靠枕做点缀；布料材质有棉麻、金属涂层、编织、皮等，常呈现做旧、磨损和褪色的效果；常用的图案有数字、字母、几何图形、建筑等。

It is recommended to select simple-shaped cushions for an industrial style space, with square and rectangle tops the list. Neutral colors of low purity are usually applied, though sometimes cushions of relatively high and pure color are used as ornaments. Fabric texture includes cotton and linen, metallic coating, woven

设计公司：PAB Architects
Design Studio: PAB Architects

工业风格空间常用抱枕
Common Cushions in Industrial Style Space

工业风格空间常用抱枕
Common Cushions in Industrial Style Space

materials, or leather, which usually creates visual effects that look old, worn, or faded. Numbers, letters (of an alphabet), geometric figures, and architectures are commonly used as the patterns.

地毯的应用在工业风格空间中也并不少见，与其他风格的空间一样，地毯也多应用于沙发区域或者床前。地毯色彩纯度较低，质地多为棉质或者亚麻编织地毯，简约的现代风格地毯与传统的波斯地毯均可用于工业风格的空间中。

It is not uncommon to see carpets in a Industrial style space. Same with the other spaces of styles, carpets are usually placed at the sofa area or in front of the bed. The carpets are in low purity colors, and mostly made of cotton and linen, or woven flax. Both modern style carpets and traditional Persian carpets are suitable for an industrial-featured space.

工业风格常用地毯
Common Carpets in Industrial Style Space

6 其他装饰品
Other Decorations

常用的工业风格装饰品有：用铁、铝等金属材质锻造的数字或字母；铁艺制品，如旧机器零件；各种类型的装饰画，最好是抽象的黑白画、字母和抽象的图案，可悬挂在墙面上，也可摆放在搁板、书架上，还可以直接置于地面靠墙摆放或者摆放在靠墙的边桌上；墙面彩绘或者在黑板漆上涂鸦也是工业风格常用的墙面装饰手法；各种机械模型也能很好地烘托工业风格气氛，如老爷车模型、蒸汽火车模型、双翼飞机模型等；交通工具，例如重型摩托车、自行车等，自行车可以直接摆放，也可以悬挂在墙上，既有装饰效果，又解决了自行车的收纳问题；带有工业时代印记的旧物，如复古风扇、老式收音机、老式拨盘电话机、机械打字机等；书籍也是工业风格的装饰品之一，可一本本叠起来直接摆放在地面上，也可置于大型金属材质的书架上。

Commonly used industrial style decorations are: Numbers or letters forged by iron, aluminum and other metal materials; Iron artifacts, such as old machine parts; All kinds of decorative paintings - good to be abstract, black and white paintings, or even better to be letters and abstract patterns - can be hung on the wall, or placed on the shelves or bookshelves, or even placed directly on the ground or on a side table by the wall; Wall painting or graffiti on the blackboard paint is also techniques that commonly used in industrial style wall decoration; Mechanical models can very good to contrast industrial style atmosphere as well, such as the classic car models, steam train models, and biplane models; Vehicles, such as heavy motorcycles or bicycles - a bicycle can be placed directly on the ground, and can also be hung on the wall, which has decorative effects and solve the problem of bicycle storage; Historical industrial old things, such as retro fans, old-fashioned radio, old dial phone, mechanical typewriters, etc .; Books are also industrial style decorations that can just be piled up on the ground, or placed on large metal bookshelves.

老式拨盘电话更像一件工艺品，带给我们不一样的回忆。摄影：Matteo Favi Immagini
An Old Dial Phone is More Like a Handicraft, Bring us Special Memories, Photographer: Matteo Favi Immagini

将自行车用挂墙支架挂在墙上，既能解放空间又能起到装饰作用。摄影：Tom Ferguson
Bicycle Hanging on the Wall by Brackets - decorating and Space Saving, Photographer: Tom Ferguson

黑色大理石，镀金属和灰色素水泥墙，让这个空间看起来冰冷而现代，复古老爷车模型作为重点装饰物摆放在黑色展台上，既为单调的素色空间增添了一抹亮色，又给空间增加了一种文艺气质

Back Marble, Metal Plating and Gray Plain Cement Wall Have Made This Space Looks Cold and Modern. The Retro Classic Car Models Were Placed in the Black Booth as a Key Ornament, Both Adding a Touch of Bright Colors and Giving the Space a Literary Temperament to the Monotonous Space.

古船木墙面装饰。工业风格强调年代感，即使不是真正的老旧脱漆的旧木板，风化木、古船木也比新板材更具视觉冲击力

Ancient Boat Wood Wall Decoration. The Industrial Style Requires Emphasizing, Thus Even If It Is not Really Old De-painted Wood or Weathered Wood, Ancient Boat Wood is also More of a Visual Impact than Some New Materials.

金属字母

Metal Letters

金属齿轮造型装饰挂钟和生锈的齿轮装饰摆件

Metal Gear Shaped Decorative Wall Clock and Rusty Gear Decoration Ornaments

抽象装饰画

Abstract Decorative Painting

复古金属飞机模型和老式机械打字机

The Retro Metal Aircraft Model and Old Machine Typewriter

第 3 章

Chapter 3

拒绝平庸的工业风格设计

The Unique Industrial Style Interior Design

塔拉萨的商用改造住宅

Commercial Space Turned into a Loft in Terrassa

项目地点：西班牙，巴塞罗那

项目面积：200m²

设计公司：Egue y Seta

设计师：Daniel Pérez, Felipe Araujo

摄影：Víctor Hugo

Location: Barcelona, Spain

Area: 200m²

Design Studio: Egue y Seta

Designer: Daniel Pérez, Felipe Araujo

Photographer: Víctor Hugo

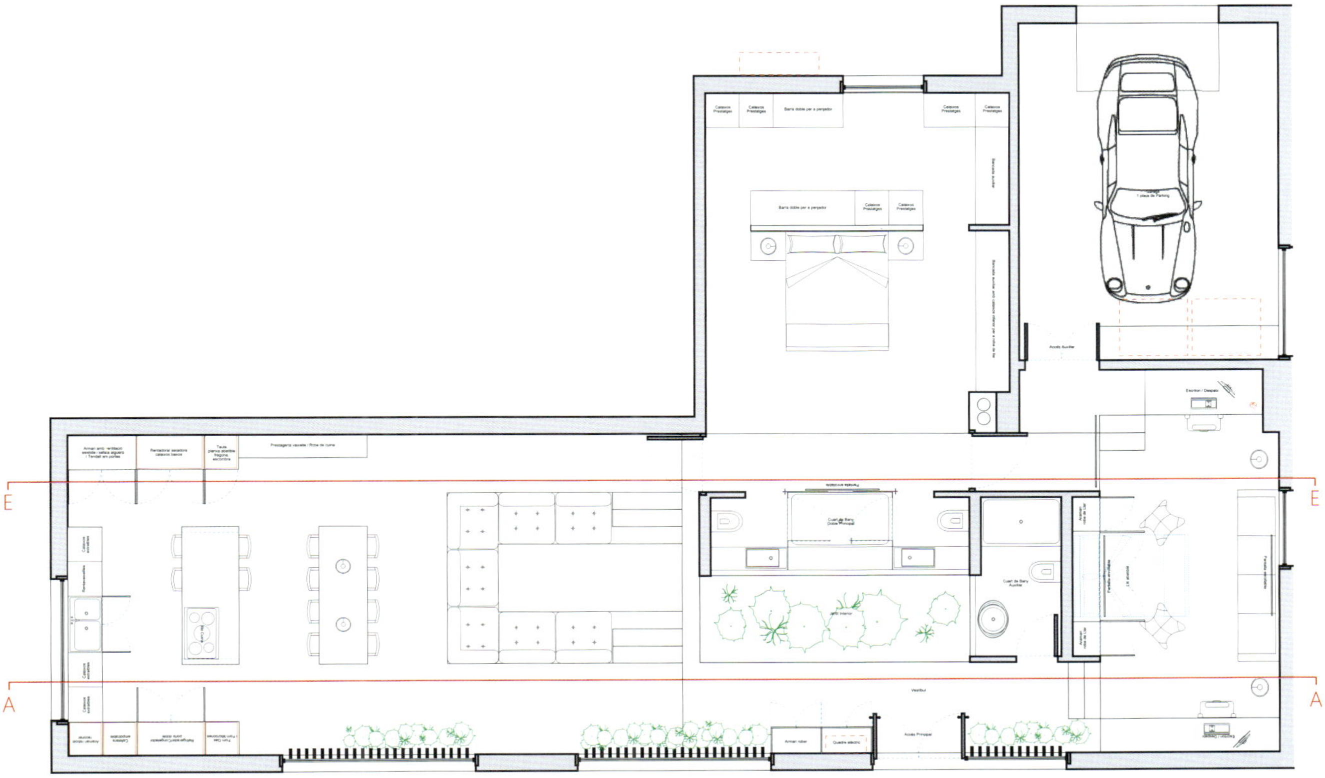

平面图

Floor Plan

Egue y Seta 没有使用最常见的空白实心墙分隔空间，而是用大片的玻璃落地窗、百叶帘以及绿色植物组成的半通透隔断来分隔，塑造出充满活力、清爽且自然的空间。设计公司通过他们最新的项目证明，只需要约 10m 长的墙面，就足以将更多的私人空间转化为与客人共享的空间。

Egue y Seta has abandoned blank solid walls to favour semi-transparent partitions made of greenery as a more sensuous and bland alternative, to make a energetic, refresh and nature space. With their latest project the studio has proven that just about 10 linear meters of walls are more than enough to articulate the more private areas of the house to those meant to welcome and share with the guests.

本案重新定义了设计师和房主口中的透明度和空间开放性。从外面看，只要把垂直百叶窗的木制叶片收拢起来，就可以看到丰富的绿色树篱和多叶灌木。要走进这昔日的商用改造公寓，我们需要打开一个宽大厚实并配有业主在自己的铸铁厂里定制的铁艺装饰的实心双扇木门。当踏上天然橡木地板时，会看到一个巨大的迎宾“鱼缸”，鱼缸里面没有鱼，只有树叶和绿色的花朵漂浮在柔和的光线中。

This house redefine what transparency and spatial openness means to designers and homeowners. From the outside, the vertical shutters draw wooden strips over a sight of a rich green hedgerow and a leafy bush. To step in this former commercial space turned into a loft, we need to open a wide and generous solid "Iroco" wood set of double doors with iron fittings custom made at the owner's iron foundry. And as soon as we set foot over the natural oak parquet, we are welcomed by a huge "fish tank" without fish, were leaves and green blossoms float in amidst almost liquid light.

客厅在入户门的左侧，沙发区域由加宽的下沉式台阶围合而成，灰色天鹅绒坐垫和靠垫排列成“U”形，当作沙发来使用。在沙发上还摆放了一些设计师设计的色彩丰富、柔和的手工复古靠垫。从客厅可以直接看到室内花园，也可以透过巨大的落地玻璃向外看。另一边是近在咫尺的用餐区和厨房，但用餐区和厨房比沙发区域的地平面要高几个台阶。

To the left, caved in the building foundations, we find the living room, surrounded by oversized sunken steps over which an "U" shaped arrangement of cement colour velvet cushions work as a

sofa. Over it, an eclectic but subtle mix of designer, hand made, and vintage cushions. From the living room area one can look directly into the interior garden, but also out, through the large window panes; and facing in the opposite direction, the dining area and the kitchen just a few centimetres away but at a higher level.

在用餐区域，精美的 Bedrock Plank 餐桌来自意大利品牌 Riva 1920，搭配的餐椅是永远不会过时的埃姆斯椅。整个用餐区由悬挂在深灰色工业天花板上的两组大型透明气球灯泡照亮。一辆由家庭工匠修复的“独一无二”的摩托车放置在右侧，突显工业风格氛围。

In the dining area, the superb Bedrock Plank A wooden table by the Italian brand Riva 1920, surrounded by the timeless Eames chairs in a concrete tone. The whole arrangement lit by two bundles of oversized clear balloon light bulbs that hang from a deep grey industrial ceiling with metallic beams. To the right a "one of a kind" motorcycle restored by the family craftsmen shows the industrial feelings.

在用餐区域后面，你会发现一个看起来非常机械化，充满工业感的厨房。由于橱柜门和侧面板是大面积闪亮的垂直不锈钢板制成，所以厨房看起来既干净又精致。同时，不锈钢板与由天然木材制成的水平台面结合，让空间在具有实用性的同时兼具舒适和温馨，并能满足主人准备健康美食和花式鸡尾酒的功能需求。

Behind the dining areas you'll find a seemingly technical and industrial kitchen that offers a clean and sophisticated look due to the large and shiny vertical planes of stainless steel that sheet all cabinet doors and side panels. But at the same time, the combination with natural wood horizontal worktops, give the space a cosy warmth while making it practical when it comes to prepare the wholesome dishes and fancy cocktails the whole area seems to be asking for.

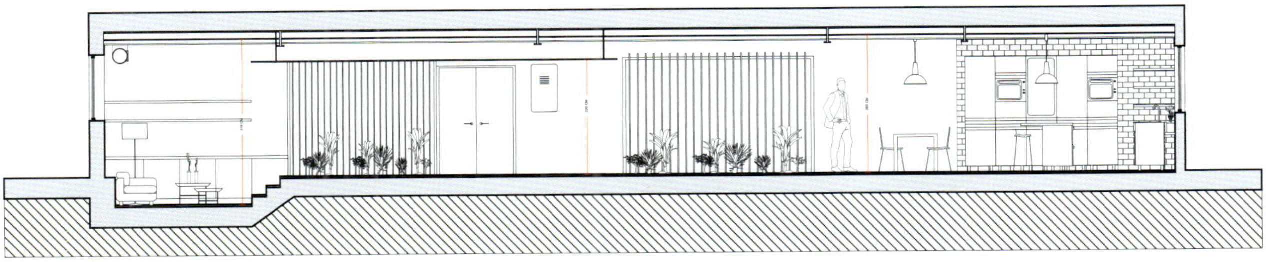

A-A 剖面图
Section A-A

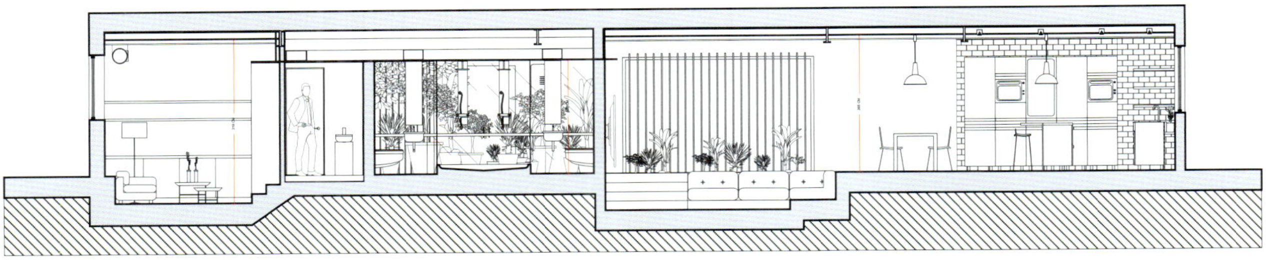

E-E 剖面图
Section E-E

主卧，设计独特的床和床头板脱颖而出。床架、收纳柜、隔板、暗藏光管的床头板和鞋架的背板，都是由天然橡木和金属格栅制成。加长的长方形浴室是对称的，可布置两个马桶和两个独立的梳妆台，中间由一个巨大的玻璃淋浴房隔开。暗藏在天花板的大型嵌入式花洒、碗状的白色洗手盆、纤细精致的水龙头和从天花板上垂下的像漂浮在绿色丛林中的镜子，以及简洁利落的悬挂式马桶、散发着光泽的白色亮面文化砖墙等元素的巧妙组合，让这个空间看起来像旧时的地铁站，呈现出一种精心搭配的都会风格。

In the master's bedroom, the bed and headboard stand out for their unique. The custom made piece fabricated with natural oak and metallic sewer grates, is at the same time bed frame, night stand, lit headboard, dividing partition and shoe rack on its back side. The elongated rectangular shape that defines the bathroom it is laid out symmetrically so to offer two toilet seats and two independent vanity tops separated by an enormous common shower tray over which an oversized ceiling mounted shower sprinkler. Over this portion, bowl shaped porcelain sinks, slim sophisticated taps and a set of mirrors hanging from the ceiling virtually floating against the green background. The suspended toilet seats and its recessed push buttons were built in walls covered with white rectangular bevelled tiles, in the style of those in old metro stations so to convey some urban feeling to the space.

穿过卧室，木地板引导我们走到房子的另一端，沿着橡木走廊走到尽头，下沉式区域再次出现，这里可以打造成一个封闭的下沉式家庭办公区和娱乐室。

Walking across the bedroom, the wooden flooring guides us to the opposite side of the house, and this vaguely varnished oak corridor, towards its end, turns itself into a desk, due to the change in the floor level that once again is caved in the foundations of the building so to create a lower and enclosed sunken home office area and entertainment room.

苏里山公寓

Surry Hills Apartment

项目地点：澳大利亚，悉尼

项目面积：300m²

设计公司：Josephine Hurley Architecture

摄影：Tom Ferguso

Location: Sydney, Australia

Area: 300m²

Design Studio: Josephine Hurley Architecture

Photographer: Tom Ferguson

本案始建于 20 世纪 20 年代，最初为一个茶叶厂，后来被改造成一个广告代理公司，现改为私人住宅。设计的出发点是为这个 300m^2 的两层私人住宅空间注入新的活力。大厦 6 楼带着标志性的拱形窗户，这里曾经被完全烧毁，从前的门卫室也被拆除了，取而代之的是装饰栏杆后面的屋顶度假公寓。

Formerly a tea factory in the 1920s and more recently an advertising agency, the brief was to breathe new life into the 300-square-metre space as a private residence that spans two levels. Level 6 with the iconic arched windows was completely gutted and the former caretaker's office was demolished and replaced with a guest retreat on the roof, located behind the emblazoned parapet.

1 Entry 入口
2 Music 音乐区
3 Living Room 客厅
4 Dining Room 餐厅
5 Kitchen 厨房
6 Study 学习区
7 Laundry 洗衣区
8 Bedroom 1 卧室 1
9 Ensuite 1 套卫 1
10 Bedroom 2 卧室 2
11 Ensuite 2 套卫 2
12 Toilet 卫生间

一层平面图
First Floor Plan

客户简要强调了舒适、实用和尊重建筑物的概念。这位客户从一个建筑师转变成一位有天分的音乐家，希望建筑作为一个背景，能够更有助于日常生活，并且因不同的使用者而做出改变。为了突出这个概念还使用了一系列内敛和低调的藏品，并且特意保留建筑物的现存遗产结构。

The client brief focused heavily on the notions of comfort, practicality and respect for the building. The client, an architect-turned-talented-musician wanted the architecture to be a backdrop that would facilitate everyday living and transform depending who was there. The concept employs a collection of restrained and understated installations that uncovered, retained and celebrated the existing heritage fabric of the building.

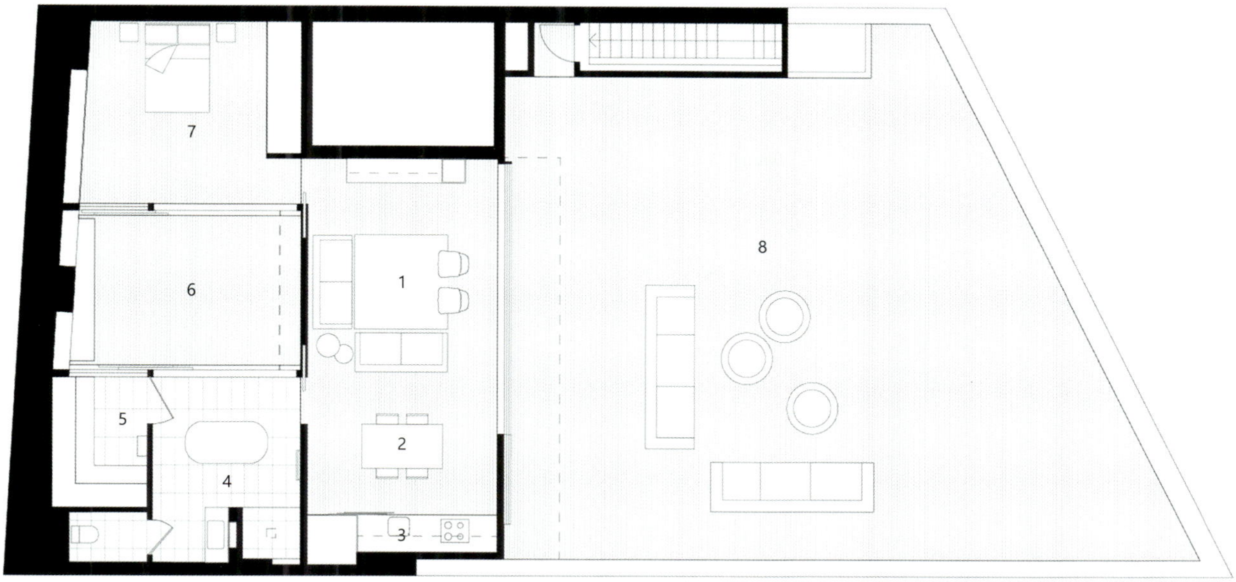

1 Living Room 起居室
2 Dining Room 餐厅
3 Kitchen 厨房
4 Bathroom 浴室
5 Sauna 桑拿房
6 Courtyard 庭院
7 Bedroom 3 卧室 3
8 Roof Deck 屋顶平台

二层平面图
Second Floor Plan

出于对住宅精致度和层高的考虑，设计师采用了暖灰色和光洁的白色这种最简单的搭配方式，既节约了成本，又兼具耐用性和实用性。本案不仅是可持续的，而且是循环再造的，充满了魅力和个性。旧建筑元素得以保留，从残破老旧到精致现代，历史的变迁让空间层次感更加丰富。改造后的公寓宁静宜居，受到了业主的欢迎。

A refined and highly-considered minimal material palette of warm grey, timber and smooth white integrated surfaces was chosen for its cost-effectiveness, but also for durability and practicality. The project is not only sustainable, it's recycled and it's full of charm and character. The warehouse is embraced and the conversion lets the layers of history enrich the interiors. The finished apartment is calm, liveable and welcoming.

Neusinger 住宅

Neusinger Residence

项目地点：美国，洛杉矶
设计公司：Juan Felipe Goldstein 设计公司
设计师：Juan Felipe Goldstein

Location: Los Angeles, USA
Design Studio: Juan Felipe Goldstein Design Inc.
Designer: Juan Felipe Goldstein

Neusinger 住宅位于洛杉矶月桂谷，设计灵感来自于法国南部的小型面包房。在这个设计中，最引人注目的元素是设计师将城市和工业设计元素融合在一起，将"旧时代"设计的产品带到一个完全现代化的结构中。

At its core, Juan Felipe Goldstein Design Inc's inspiration for the Neusinger Residence in the Laurel Canyon area of Los Angeles was a small bakery in the south of France. Perhaps the most striking element of Goldstein's design, however, was his integration of urban and industrial design elements to bring a quality of "old world design" to an entirely modern structure.

设计师利用现代化的视角，对这个中世纪之家进行了修缮和扩建，在原结构的基础上面积增加了 372m²，主建筑改造成 4 层高 5 间卧室的现代结构住宅，创造出一个流动的、开放的立体循环空间。每个空间都可以通过水平和垂直两个循环系统连通为一体，又各自独立。通过大面积的玻璃窗，全年都可以看到月桂谷景观和周边自然环境。与主体建筑相连的是一个新的工作室和一个可以作为独立家庭工作室或客房的附属公寓。

The renovation and addition to the mid-century home retains elements of the original structure but through an entirely modern lens. Adding $372m^2$ and transforming the main house into a 4-level, 5 bedroom modern structure, a fluid, open circulation was created throughout, each space given its own boundary while connecting to the larger house through both vertical and horizontal circulation. Distant and intimate views were created through an abundant use of windows and nature was brought indoors year round to take advantage of the Laurel Canyon landscape. Attached to the main structure is a new office and an additional loft for a separate home office or guest apartment.

客户希望自己的房子带有法国南部风情，设计师从一座他曾经在法国南部见过的砖砌面包房中获取灵感，小扁豆、横梁和地道的乡土建筑中的常见元素，被无所顾忌地展现出来。在把一个旧的主浴室改造成餐厅时，设计师让工人把瓷砖全部拆除，露出一面混凝土墙，最后把空气管道、抛光的混凝土地板等全暴露出来。设计的最终目的是要打造一座真实的建筑，通过一个完全现代的镜头可以看到基本的、粗略的结构。

To satisfy the client's desire for a house reminiscent of the south of France, Goldstein took his inspiration from an old brick bakery he'd seen once in France with exposed lentils and beams and the kind of unashamed, authentic elements common to vernacular architecture. "When we transformed an old master bath into the dining room," says Goldstein, "I had the workers chip away at the tiles to reveal a raw concrete wall, and throughout the house, we exposed air ducts, polished concrete floors, and so forth. I wanted to create an architecture without modesty where basic, provincial structures are visible through an entirely modern lens."

客户要求房子要具备防火功能，尽量使用钢和砖为建材，所以设计师设计了一种特制的小混凝土砖块，以达到与原建筑旧砖块契合的效果，所有的墙都用这些砖块砌成。新增的建筑全部由砖和钢建造，不仅强调材料的可持续性，还通过利用太阳能资源，设计出靠吸收太阳辐射采暖的混凝土地板，形成全屋绿色循环系统。

The client also requested that the house be steel and brick to be as fire-proof as possible, so Goldstein designed small, specially-made concrete block bricks and all walls are made of the brick to match the old brick of the original house. The addition was made of brick and steel, with an emphasis on sustainability - making it energy efficient with passive solar strategies, concrete floors with radiant heating, and green finishes throughout.

波斯的阳光小屋

House of the Sun in Persian

项目地点：乌克兰，基辅

项目面积：101m²

设计公司：Keivani Architects & Persian Primavera

设计师：Nima Keivani, Sina Keivani, Solmaz Fooladi

摄影：Sergey Savchenko

Location: Kiev, Ukraine

Area: 101m²

Design Studio: Keivani Architects & Persian Primavera

Designer: Nima Keivani, Sina Keivani, Solmaz Fooladi

Photographer: Sergey Savchenko

本案是住宅区的一间平层公寓。设计师通过两个必要的因素来体现客户的需求：第一个因素是客户对波斯的文明、文化和价值观很感兴趣，即对密特拉神和太阳的崇拜；第二个因素是客户的艺术个性，需要在设计中反映出来。考虑到预算有限，我们的目标是打造一个虽小但具有密特拉教艺术和伊朗特色的空间。

The project is located in one of the apartment block of residential complex. During the design, we considered two essential factors to develop the need of owner. The first factor is the client's interests in the Persian civilization, culture, and values, namely the metaphors of Mithraism and sun. The second factor is the artistic character of the client, which had to be reflected in design. Considering the budget restrictions, we aimed to develop a art space that owner can afford the pay and affordable space, yet artistic and Iranian genius loci with a symbolic approach to Mithraism.

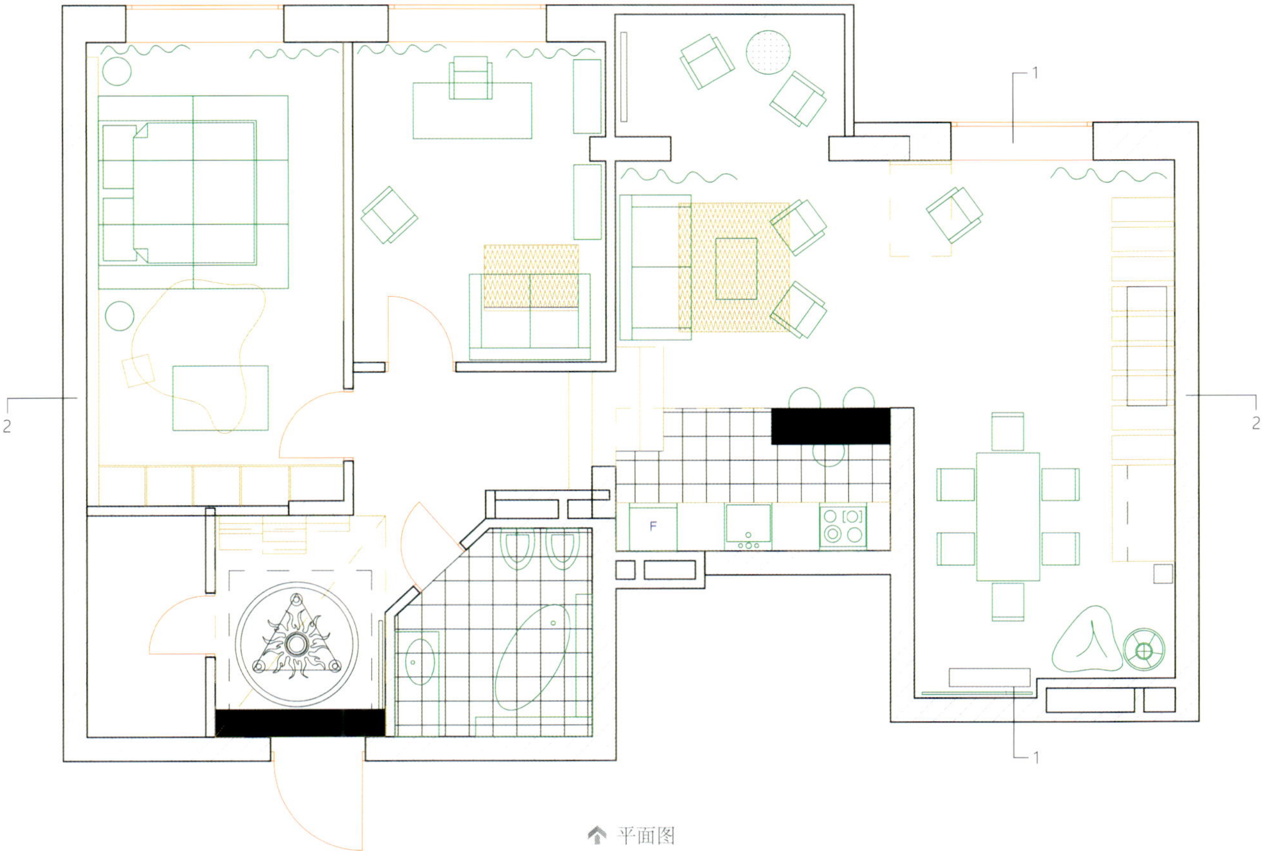

平面图
Floor Plan

设计师采用清水混凝土、钢筋和金属氧化物来设计空间。选择砖和木材，以及暖光照的设计创造了一个舒适和温暖的气氛，而这正是一个住宅所需求的。

Designers used exposed concrete, rebar, and metal oxide to design the space. The choice of brick and wood as well as lighting design created a cozy and warm atmosphere that is desirable for a residential place.

恰高·占比尔神殿
Chogha Zanbil Ziggurat

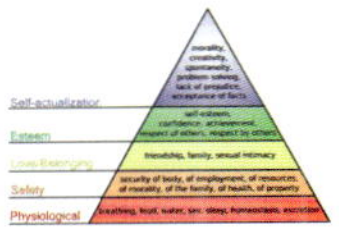

马斯洛的需求层次理论
Maslow's Hierarchy of Needs

光明战胜黑暗
Light Overcome The Darkness

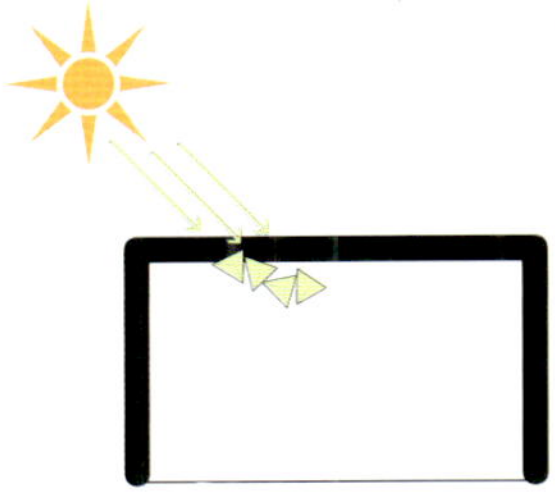

光线穿过混凝土天花板
Light Passing Through The Concrete Celling

设计的重点之一是位于客厅的楼梯，它是一个具有象征性的哲学元素。这个设计象征接近太阳，体现了艺术家的思想，也寓意进步的阶梯和对知识的追求。楼梯的上升方向向东，是日出的方向。

One of the most important design ideas is the set of stairs located in the living room; it is a symbolic, yet philosophical element. This design is a symbol of approaching to sun and advancing to knowledge essential for improvement and actualization of thoughts for an artist. The direction and raise of stairs are toward east and sunrise.

照明灯具的设计也受到太阳的启发，太阳的多种几何符号里，有一种由十二个三角形包围而形成太阳符号的图案，常用于伊朗传统建筑。三角形和金字塔形灯具的使用，使它们既是光的载体，也是光的延伸。

The design of lighting fixtures is also inspired by the symbolic role of Sun, and it follows the geometric symbol of sun, which a triangle surrounded by a dodecagon, used in traditional Iranian architecture. In a part of ceiling the triangular and pyramidal shapes are used such that they are such as the extensions of the light radiation that pass through the roof and shines inside.

围绕古太阳的设计元素，家具设计、厨房装饰设计等都延用这一概念。家具通过使用混凝土、钢筋和金属材料来传递对太阳象征意义的认识。设计师还在这个空间中利用了绿松石瓷砖，以激发伊朗传统建筑中的场所感。

The ancient sun pattern was used as a light frame in this unit. Furniture are designed in accord with the aforementioned concepts, realized by using materials such as concrete, rebar, and metals. Not only designers applied this concept to the interior design of kitchen for consistency, we also used turquoise tiles in this space to inspire the sense of place that emerges in the Iranian traditional architecture.

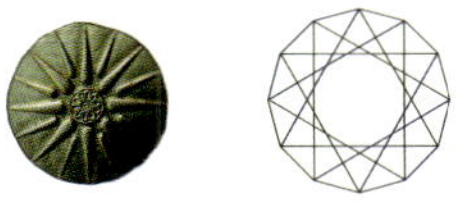

波斯古太阳图案
Persian Ancient Sun Symbol

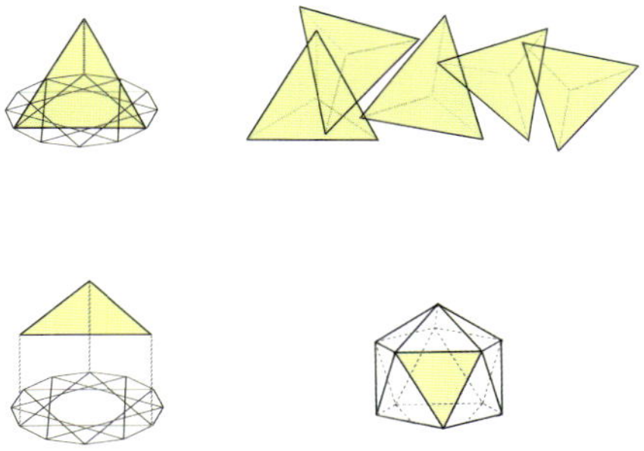

灯具造型设计过程
Lighting Fixture Design Process

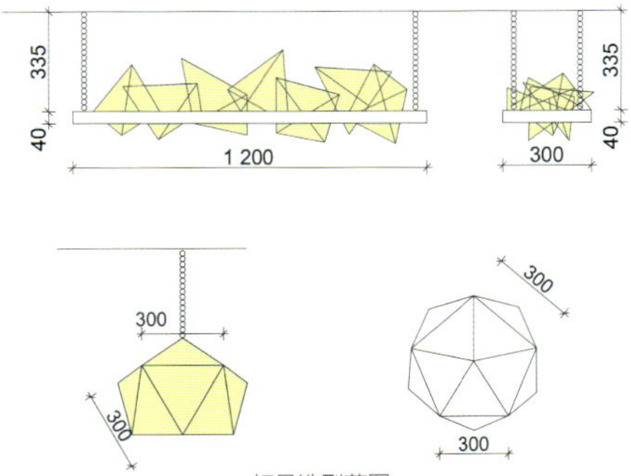

灯具造型草图
Lighting Fixture Drafting

尽管在卧室的设计中使用了诸如混凝土、钢筋和回收木材等经济实惠的材料，但设计师创造出了一个理想和轻松的空间，以满足卧室的需求。

Although affordable materials such as concrete, rebar, and recycled wood are used in the bedroom's interior design, our design has created a desirable and relaxing space to satisfy the requirements of a bedroom.

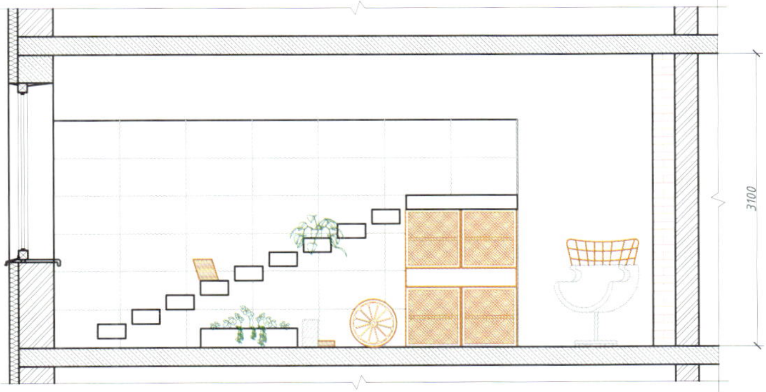

1-1 剖面图

Section 1-1

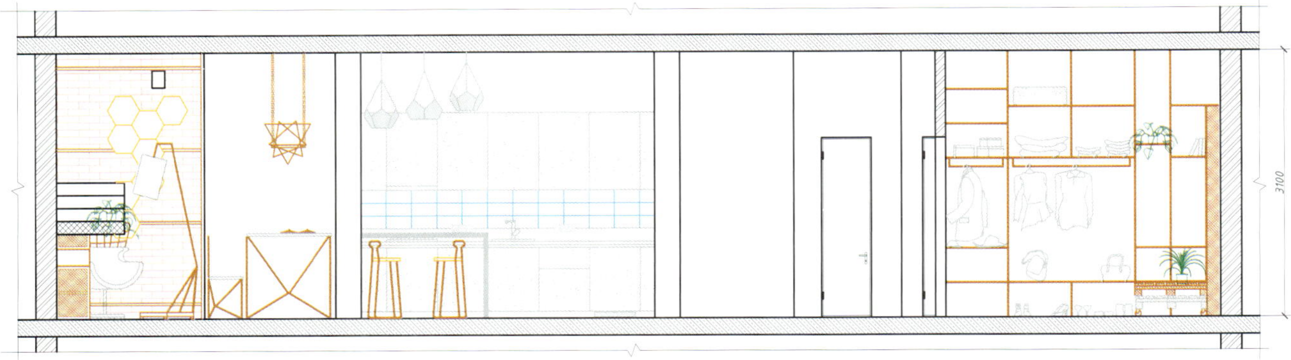

2-2 剖面图

Section 2-2

响乐空间

The Sounding Space

项目地点：立陶宛，维尔纽斯

项目面积：120m²

设计公司：PRUSTA LTD ARCHITECTURE & CONSTRUCTION

设计师：Ieva Prunskaite

摄影：Leonas Garbaciauskas

Location: Vilnius, Lithuania

Area: 120m²

Design Studio: PRUSTA LTD ARCHITECTURE & CONSTRUCTION

Designer: Ieva Prunskaite

Photographer: Leonas Garbaciauskas

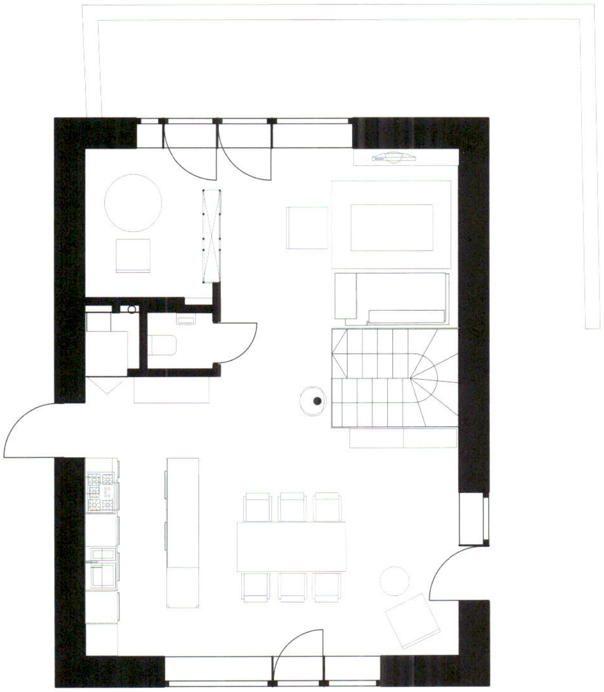

这房子宽敞的内部和谐地融合了精美的内饰焦点：开放的整体建设和天然灰木形成的生动的东方背景。白色大理石、黑色金属、玻璃、不锈钢、软纺织刀片、老化木材，这里一切都是活着的、自然的、非常真实的。

This spacious interior of this house harmoniously combines delicious interior highlights: open monolithic construction and natural greyish wood become a beautiful background for the vivid oriental pieces. White marble, black metal, glass, stainless steel, soft textile inserts, aged wood - here everything is alive, natural and very real.

↑ 二层平面图
Second Floor Plan

这个四口之家位于维尔纽斯的居民区，周围是美丽而宁静的森林。设计的目标是创造风格简洁的空间，并和现有的室内元素相匹配，如旧的柜子、桌子、椅子和地毯等。因为这些家具的颜色都很鲜艳，而且以暖色调为主，所以墙壁、天花板和地板的颜色应该尽可能地选择中性色调或冷色调。

The four-person family live in the residential district of Vilnius which is surrounded by the remarkable beauty and tranquillity of the forest. One of the goals for the new interior was to create stylistically uncluttered space that is suitable for already existing furniture peaces and interior elements such as old cupboards, tables, chairs and carpets. As those things are quite vivid and warm colours dominant, so walls, ceilings and floor colours should have been as neutral as possible and also in cool shades.

本案的室内设计理念不是独立的各种设计元素，而是各种元素之间的关系和相互作用。极简主义的墙壁、天花板和地板的设计重点是突出一些局部设计。不同的材质、纹理之间的对话是在这里创建的，例如木材和混凝土之间，石头（意大利石材）和木材之间，白色大理石和混凝土之间。

The main principal of this house interior concept is not the separate elements, but rather their interaction: minimalist walls, ceiling and floor solutions are such that highlights some specific interior accents. So, first of all, dialogues between different textures such as wood and concrete, stone (Italian slate) and wood, white marble and concrete are created here.

家居设计的诱人之处在于，随着空间完成度的提高，业主的爱好和兴趣也随之展现。厨房区域可能是反映业主爱好的主要空间，装备齐全的厨房让人感觉可以品尝到媲美餐厅的美味，这里使用了功能强大的工业厨房设备，比标准的燃气灶具、钢制橱柜和工作台更大，这样的厨具既功能强大又干净易维护。

Creating the interior for a family housing, it is always tempting to reveal hobbies and interest of the people who will live here. Probably, kitchen area is the space that reflects owners hobbies most of all. It recalls the real restaurant's cuisine: the powerful industrial hood is used here, larger than a standard gas cooker, steel cabinets and worktops. Such furniture is very functional - it is hygienic and easy to maintain.

业主的另一个爱好是旅行。从在不同空间展示出来的纺织物中可以发现，大多数的地毯和挂毯是业主在非洲、亚洲等地区的国家旅游时带回来的。

Another hobby of the inhabitants of this house owner is - travelling. It is quite distinctly revealed by the textile elements that are exhibited in different spaces the house. Most carpets and tapestries were brought from travelling in Africa, Asia and other countries.

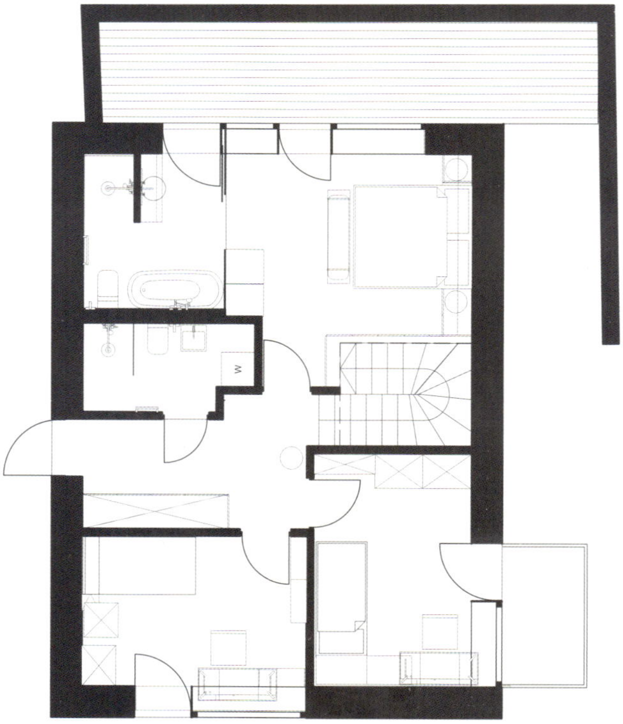

一层平面图

First Floor Plan

爱国东路王公馆

Aiguo E. Rd.Wang Mansion

项目地点：中国，台湾省，台北市

项目面积：181m^2

设计公司：奇拓室内装修设计有限公司

设计师：Chlo'e Kao, Ciro Liu

Location： Taipei City, Taiwan, China

Space area：181m^2

Design team：CHI-TORCH Interior Design

Designer：Chlo'e Kao, Ciro Liu

本案是一栋有 40 年楼龄的老公寓，为尊重老房子原有的历史痕迹，设计公司刻意保留有年代感的红砖，来彰显历史的价值。墙面造型延续复古的概念，材料选用旧木料，裸露出通常会被掩饰的建筑材料，在强调历史感的同时也满足使用者的生活需求。

It is a case about 40 year-old vintage apartment. To respect its history, CHI-TORCH deliberately made conscious efforts to retain details of time manifestation and show the value of it. Wall shape continues the concept of respecting history, accessing the Taiwan's unique old timber as its material, which can displaying building materials that many try to conceal, and satisfied the lives of the users.

设计师将极简主义和工业风格融入空间中。通过降低窗台高度、加大窗户面积的方法，让窗外的美景、光线与室内产生对话。另外，室内还摆放了再回收利用的铁艺装饰与有历史感的旧家具，如：定制铁件书架、跳色人字拼地板、旧工厂回收的吊灯、餐椅等。

The concept is blended with minimalism and industrial. CHI-TORCH specifically seeks out prefect mixture of sunshine and inner space by lowering window rails to maximize outside picturesque scene. Besides, recyclable iron pieces and historical old furniture were played in the inner space, including customized iron bookcase, chandeliers recycling from old factory, dining chairs and so on.

总的来说，设计者尽可能地兼顾空间中对风格和功能上的要求，使复古元素与现代元素和谐地融入这个空间中。

To sum, CHI-TORCH passionately designed and selected pieces that are as much about function as style to add historic and modern co-existing spirits to the most thoughtfully designed home.

霍兰德酒店

The Hollander Hotel

项目地点：美国，芝加哥

设计公司：DELORDINAIRE

摄影：Adrian Gaut

Location: Chicago, USA

Design Studio: DELORDINAIRE

Photographer: Adrian Gaut

本案位于充满活力的柳条公园，与巴克镇社区为邻，在芝加哥市中心环路和罗根广场之间，坐落在建成于 1905 年、有五层砖结构的霍兰德防火仓库，并与它的姐妹酒店罗比酒店毗邻。这个由法国设计工作室 DELORDINAIRE 设计的高端场所配备了 66 张床，包括 12 间住房、可以供应新鲜咖啡的咖啡厅、私人储物柜、一楼的洗衣设施和自行车租赁及维修店。小型自行车租赁店和咖啡厅并列在一个 8.2m 长的木制工作台上。客人自助洗衣店与接待处的大长皮椅并列，在这里客人可以在洗衣的同时享用咖啡。

Located in the vibrant Wicker Park and Bucktown neighborhoods between downtown Chicago's Loop and Logan Square, The Hollander sits in the historic 1905 five-story brick Hollander Fireproof Warehouse adjacent to its sister property, The Robey. This high-design Social Stay from French design studios DELORDINAIRE is equipped with 66 beds, including 12 private rooms, a café serving freshly brewed coffee, personal lockers, ground floor laundry facilities and an onsite bike rental and repair shop. A small bike hire shop and a café sit side-by-side on a 8.2 meters long wood workbench. The guest laundromat is juxtaposed with the reception with large leather banquettes where guests can enjoy a coffee while catching up on their laundry.

原建筑是一座古老的防火仓库，是芝加哥工业化历史的见证。设计师通过保存和显露现有的结构和特征，展示出建筑的实用性。材料和配件延续这一设计理念，管道裸露在外，霓虹灯点亮走廊，大堂装修由混凝土镶板和桦木胶合板组合而成，卧室里混凝土地板和天花板也暴露在外，床架由桦木胶合板和碳钢组成，搁架是由折叠钢定制而成。为了把关注点放在优雅而实用的材料上，设计公司定制了全屋的家具和照明设备。

The original building is an old fireproof warehouse: a testament to Chicago's gritty industrial past. Preserving and exposing the existing structure and character, DELORDINAIRE architects celebrate the utilitarian nature of the building. The materials and fittings continue the design concept; conduit is left exposed and neon lights punctuate the corridors. The lobby finishes combine concrete panelling and birch plywood. In the bedrooms, the existing concrete floors and ceilings are left exposed. The bed frames, are built in birch plywood and black steel. Shelving is custom designed from folded steel. With an accent on elegant yet utilitarian materials, DELORDINAIRE custom designed the furniture and lighting throughout.

疯狂巨人酒吧

Mad Giant Beer Bar

项目地点：南非，约翰内斯堡

设计公司：Haldane Martin Iconic Design

设计师：Haldane Martin

摄影：Micky Hoyle

Location: Johannesburg, South Africa

Design Studio: Haldane Martin Iconic Design

Designer: Haldane Martin

Photographer: Micky Hoyle

约翰内斯堡市中心的这个啤酒餐厅旨在体现疯狂巨人的意识形态，并创造一个空间，内部尺寸参考了大号的金属组装玩具。在这里，顾客不仅可以体验到自由自在、充满冒险精神的品牌哲学，同时还可以在充满活力的环境中体验到独特的风味和无拘无束的创意氛围。

The brewery and restaurant in downtown Johannesburg was designed to embody the Mad Giant ideology which plays with scale referencing oversized metal toy construction kits in the interior space, and to create a space where beer drinkers can experience the free-spirited, adventurous philosophy of the brand, while experiencing a celebration of flavour and uninhibited creativity in a vibrant environment.

直径 6 米的环形混凝土吧台像巨人脚下踩的巨型瓶盖。啤酒厂背面还有三个小型的混凝土酒吧，所有的家具都是定制的，包括由 Haldane Martin 设计的一系列将原创钢结构玩具放大变成的家具，将品牌理念贯彻其中。墙壁上有涂鸦艺术家 Nomad 创作的巨幅灰色肖像壁画，俏皮而童真，强调了孩童般的远大梦想为生活带来的活力与动力。

The 6-metre-diameter cast concrete bar counters take the form of a giant bottle cap beneath the giant yeti. There are another three simpler concrete bars at the back of the brewery in the tasting pods. All the furniture in the brewery is bespoke and includes a range of original pieces by Haldane Martin made from custommade scaled up Meccano, like giant toys turned into furnishings, to bring the brand vision to life. Giant murals with a playful, childlike aesthetic adorn the walls, enhancing the effect of a child's big dreams brought to life. The walls of the brewery feature grey-scale portrait murals by graffiti artist Nomad.

地板用不同材料拼贴而成，像放大的玩具拼图。用餐区的地板选择的是回收的罗得西亚柚木镶木地板，铺成人字形。巨型瓶盖形状吧台前的地板看起来就像像素化的碎片，深色木地板碎片与混凝土融合形成俏皮的图案。餐椅的扶手和前腿都是由钢筋制成的，支撑靠背的钢结构椅腿被喷成了明亮的红色，在靠背处用激光切割出一个疯狂巨人标记，坐垫和座椅靠背由仿旧皮革制成。

The floors combine different materials and reiterate the trope of blown-up toy building blocks. The dining area floor is made from recycled Rhodesian teak parquet, laid in a herringbone formation. At the front of the brewery the herringbone fragments into pixilated pieces, merging into stained concrete with floating pieces of wood creating playful patterns. The dining chairs have rebar arms and legs, metal back and legs sprayed bright red with a Mad Giant emblem laser-cut into the back, with seats and seatbacks upholstered in distressed leather with red stitching.

卫生间洗手台造型像“工”字形的横梁，由漆成亮红色的结实的金属制成。卫生间和厨房都贴了蜂巢状的瓷砖，厨房甚至用明亮的红色瓷砖水泥浆勾缝，以增强视觉效果。

The bathroom has massive I-beams painted bright red converted into basins. The hexagonal pattern of the bottle store floor tiles is continued in the bathroom's scaled-down honeycomb wall tiles, as well as in the kitchen, emphasised with bright red tile grout.

NO PARKING

MAD
GIANT
BOTTLE
SHOP

墨西哥 Havre 77 商住多用住宅改造

Havre 77 Intervention, Mexico by Francisco Pardo Arquitecto

项目地点：墨西哥，墨西哥城

项目面积：1485m²

设计公司：Francisco Pardo Arquitecto, Julio Amezcua

设计师：Francisco Pardo Arquitecto, Julio Amezcua

Location: Mexico City, Mexico

Area: 1485m²

Design Studio: Francisco Pardo Arquitecto in collaboration with Julio Amezcua

Designer: Francisco Pardo Arquitecto, Julio Amezcua

这个根据其所在地命名的建筑由 Francisco Pardo Arquitecto 和建筑师 Julio Amezcua 合作设计，Havre 77 是一个城市空间再激活再利用项目。本案位于标志性的墨西哥城改革大道南侧，也是 Juárez 殖民地区改造的一部分。如今繁忙的市中心在 20 世纪早期曾是该市知名的郊区之一，在经历了革命战争和两次摧毁性的地震之后，这里经历了将近 50 多年的萧条期。

As its name which reflects its exact address would suggest, Havre 77 by Francisco Pardo Arquitecto in collaboration with architect Julio Amezcua is an urban re-densification and reactivation project. Standing on the south side of the emblematic Paseo de la Reforma, the intervention is part of a wider regeneration program covering Colonia Juárez. Today a bustling central district, the area used to be one of the city's most exclusive suburbs back in the early 1900's, before it was hit by a revolutionary war and two destructive earthquakes in 1957 and 1985, which led to a rent freeze for over 50 years.

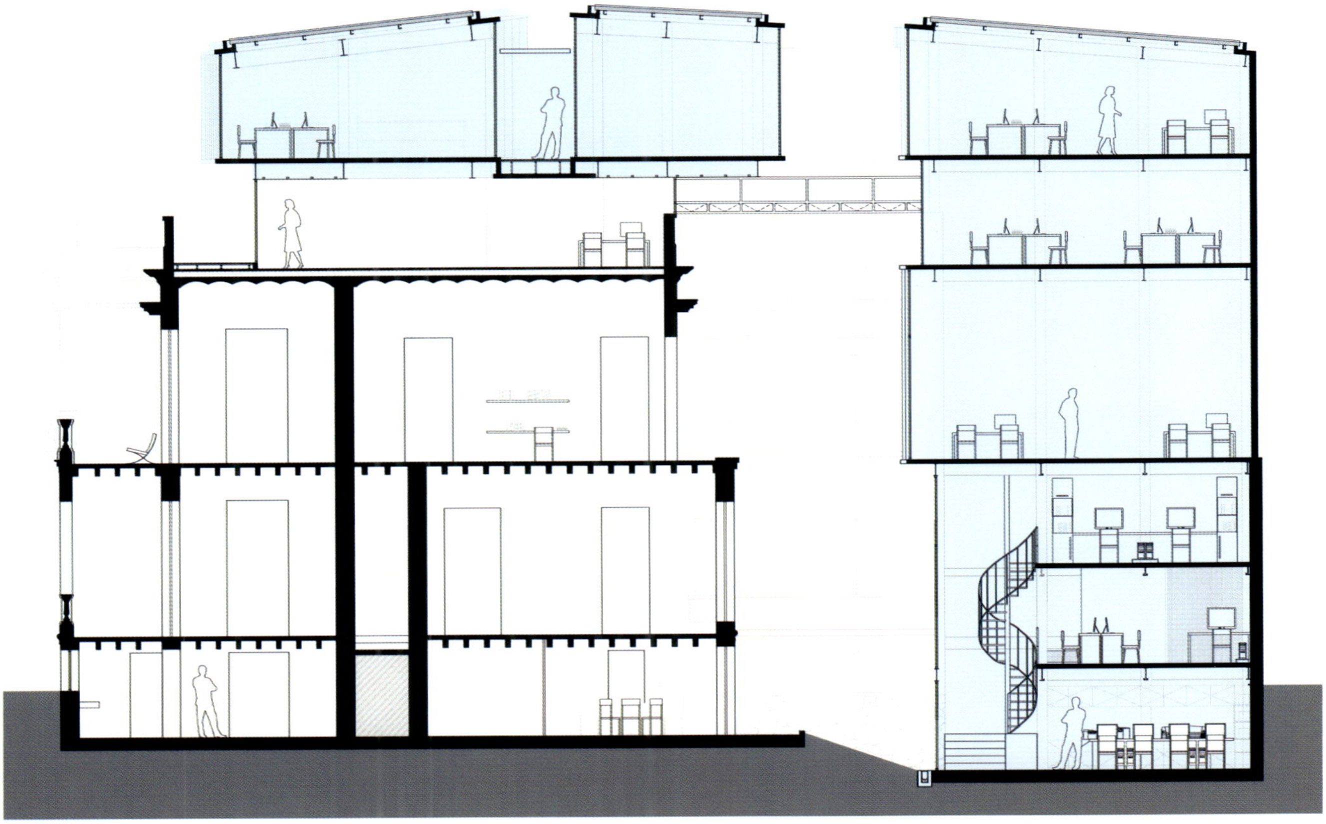

立面结构图

Elevation Structure

这栋建于十九世纪的建筑曾经居住着一户中产阶级家庭，现已被彻底改为混合功能空间，包括办公室、共享工作空间、一个法式餐厅和一个日式餐厅。新增建的两处钢筋混凝土结构建筑，像人体的假肢一样，一处安置于现存砖砌建筑的楼顶，另一个处位于后面。通过不同的建筑语言唤起不同的时代，现存的建筑和新的建筑自然地融合在一起，并像一个硬币的正反面一样完整。

A 19th century house, once accommodating an upper middle class family, has been radically transformed into a mixed-use venue including offices, co-working spaces, and a French and a Japanese restaurant. Like the prosthesis of a human body, a new steel and concrete structure comprising two floors grips the top of the former brick building, while an additional branch lies at the back of it. While evoking different eras through different languages, the existing building and the new additions are naturally integrated and complete each other like two sides of the same coin.

WS 2
WS 1

H77
3N FONDEADORA
2N MERO MOLE
RECO 77
ALMA & MATER
HAVRE 77
KVD

这个项目旨在改变该街区的 DNA 以适应现代社会的需求，Harve 77 向这个曾被废弃的街区注入了新鲜的血液，其面对大街开放，可通过南面广场的间隙到达。本案加强了建筑与周边街区的内在联系，并有助于修复街区残破的城市面貌。

This project aims to change the DNA of a neighborhood to respond to current social needs. Instilling new life-blood into an abandoned lot, Havre 77 opens onto to the street and brings it in, through an interstitial plaza on the south side. In this way, the project enhances its intrinsic connection with the surroundings and contributes to healing the neighborhood's fractured urban texture.

↑ 法式餐厅
French Restaurant

↑ 日式餐厅

Japanese Restaurant

居住空间
Living Space

BECYCLE 精品健身房

BECYCLE Boutique Fitness Studio

项目地点：德国，柏林

项目面积：420m²

设计公司：Lien Tran Interior Design

摄影：Sven Philipp, Waldemar Salesski

Location: Berlin, Germany

Area: 420m²

Design Studio: Lien Tran Interior Design

Photographer: Sven Philipp, Waldemar Salesski

BECYCLE 的主体装饰是一个复杂的几何结构，由各种不同规格的面板制成，表面覆盖拉丝黄铜，这样的组合形成了炫目的、独立的、动态的空间配置，与暴露出来的建筑结构和粗糙的砖砌墙面，形成巨大的视觉反差，从而创造出老旧与现代并存的空间。这种精致炫目的几何结构装饰从工作室的入口开始，以“展示墙”的形式出现，体育用品和配件展示在各种尺寸的壁龛中。随着动线的推移，它逐渐扩展成一个三维的立体，并逐渐融入浴室和工作室中。

BECYCLE's centrepiece is a complex geometric structure made from a variety of different panels with a surface of genuine brushed brass. This assembly establishes a dazzling, independent, dynamic spatial configuration that contrasts with the building's existing exposed architecture and raw brickwork to createa space where heritage meets the cutting edge. The structure begins at the entrance to the studio, where it acts as a "retail wall" with sporting goods and accessories displayed in niches of various sizes. Moving towards the centre of the studio, it expandsto form a three-dimensional body that incorporates the entrances to the bathroom facilities and staff rooms.

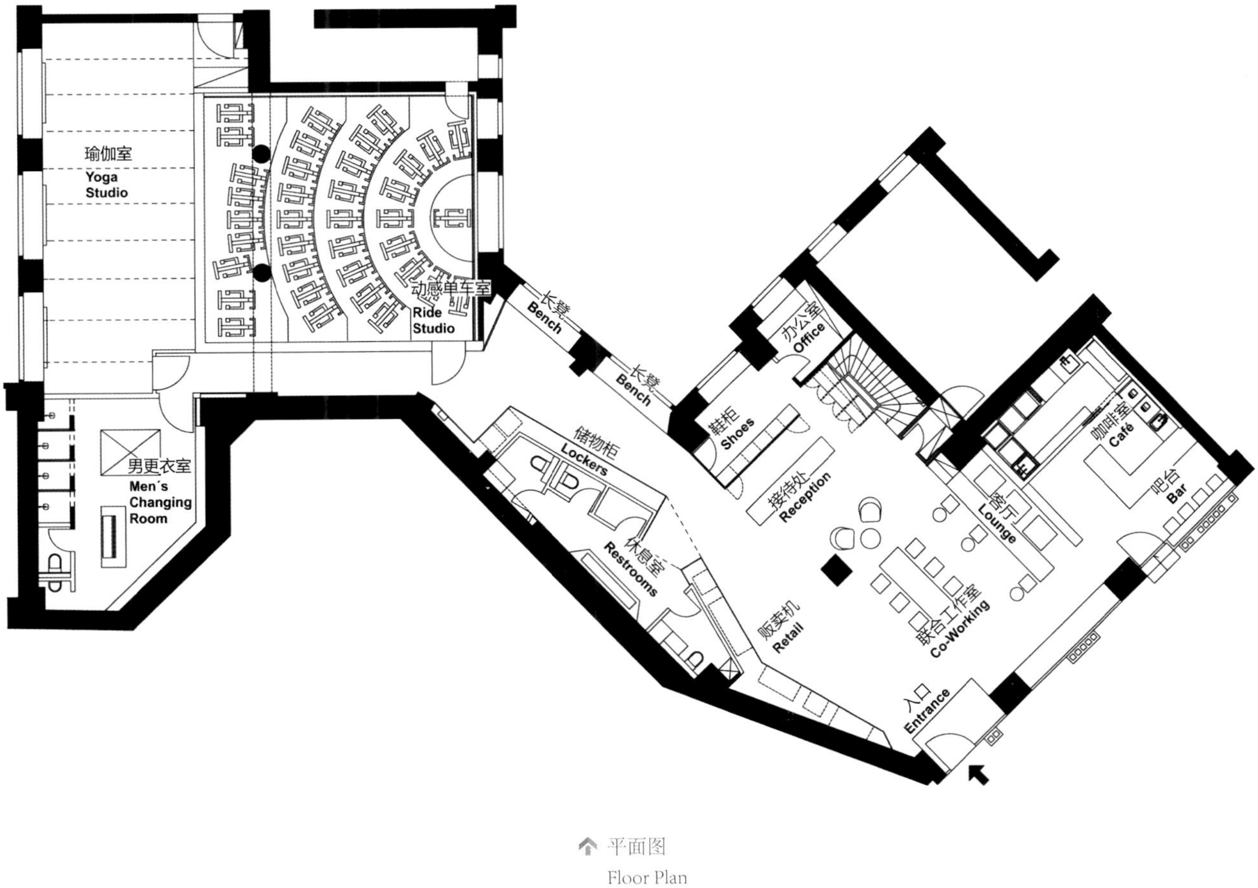

平面图
Floor Plan

BE CYCLE

SHOES
TOWELS

MADE.COM

之前作为银行地下室的原有建筑的拱顶与墙壁，已经被转换为安置女子淋浴设施。男士的淋浴间从天花板一直到地板都用石板瓦平铺而成，使每个隔间像凿出来的一块岩石。

The vault in the cellar – arelic from the good old days when the building served as a bank – has been converted to house the women's shower facilities. The men get to savour another kind of aquatic experience: their showers are fitted floor-to-ceiling with slate tiles, making the doorways to the individual cubicles look hewn out of a slab of rock.

STAGES

兔子洞——有机茶室

The Rabbit Hole - Organic Tea Bar

项目地点：澳大利亚，悉尼

设计公司：MATT WOODS DESIGN

设计师：Matt Woods

摄影：Dave Wheeler

Location: SYDNEY, AUSTRALIA

Design Studio: MATT WOODS DESIGN

Designer: Matt Woods

Photographer: Dave Wheeler

本案彻底颠覆了传统、陈腐的茶室设计概念。设计师保留了原工业建筑的结构，抛光的混凝土地板、挑高的人字形木结构天花、裸露的砖墙。通过清洁刷白暴露出的建筑元素，扩大东北向的窗户，使大量柔和的自然光线进入建筑内部来，让原本粗粝硬朗的风格更加柔和了，同时也保留了工业建筑的个性。

The Rabbit Hole - Organic Tea Bar, reinvents the tired & clichéd teahouse concept. The design begins by taking advantage of the former industrial sites inherent architecture. Concrete floors are polished; herringbone strutted timber ceilings are unearthed; original brick walls are revealed. The softening of this masculine architecture is achieved through the white washing of these newly exposed elements, & through the addition of enlarged northeast facing windows, which allow light to flood in to the interior.

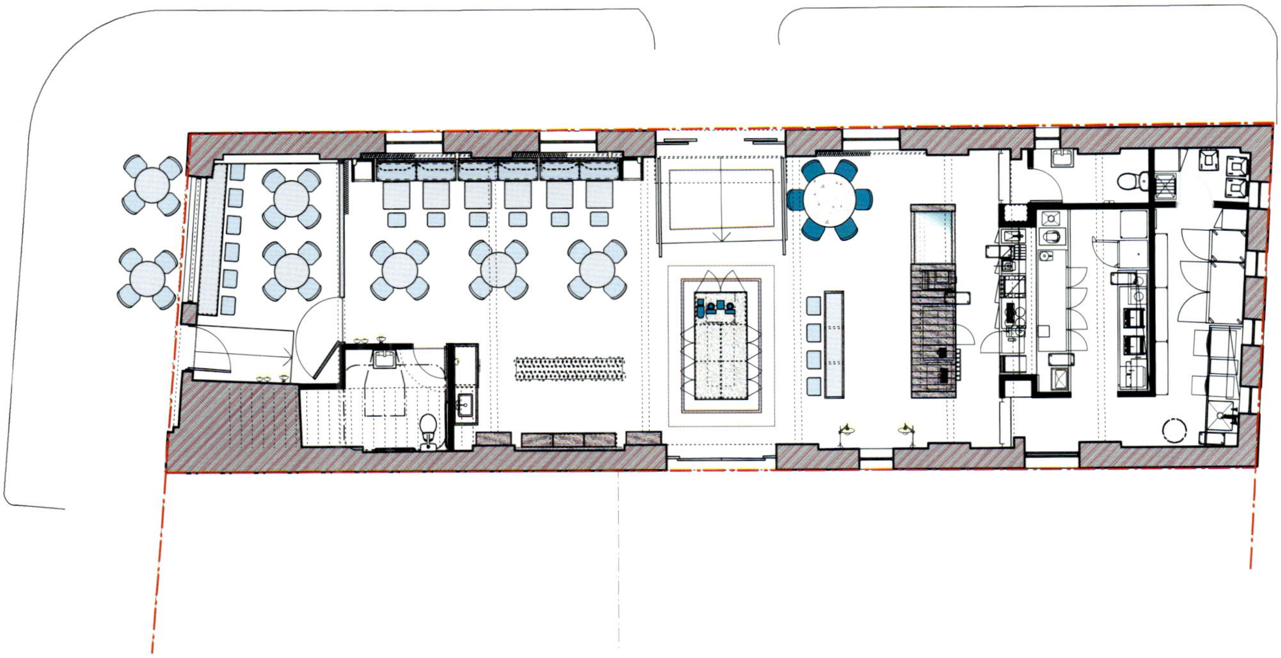

平面图

Foor Plan

“金缮”艺术（运用纯天然材质修补残缺器物的工艺名称）奠定了设计元素的基础。这最能体现在一组名为“茶具展示”的艺术装置中，一些定制的金缮瓷碗安置在橡木圆柱顶端，像马戏团的转盘表演。金缮概念还通过大面积碎裂陶瓷的使用来重现，这些碎裂的陶瓷在咖啡厅主服务区的零售展台上独显特色，在这个陶瓷覆盖的巨大展台上，悬挂了一盏由茶包（茶包由智利艺术家 Valeria Burgoa 设计）制成的枝形吊灯。再生橡木做成餐厅座椅和桌面框架，坐垫和有黄铜搭扣的靠垫由皮革和棉麻纺织物组合而成，服务台由历史悠久的法式橡木地板龙骨制成。

The art of Kintsugi （which means to join with gold, and is a celebration of the beauty of imperfection in ceramic objects） forms the foundation of the new design elements. This is most apparent in the "Specialty Tea Display" where, custom designed Kintsugi bowls sit delicately above turned oak timbers，like spinning plates on top of a circus performers pole. The Kintsugi concept is further reinforced with the expansive use of a crushed ceramic tile, which is featured on the service-dominated areas of the café such as the central monolithic retail display. And above this tile clad monolith, a chandelier made entirely of tea bags （by Chilean artist Valeria Burgoa） has been installed. Reclaimed oak timbers create banquette seats & table frames. A combination of leathers & upholstery fabrics create seat cushions & brass buckled back rests. Century old French oak floor joists are used to create the service counter.

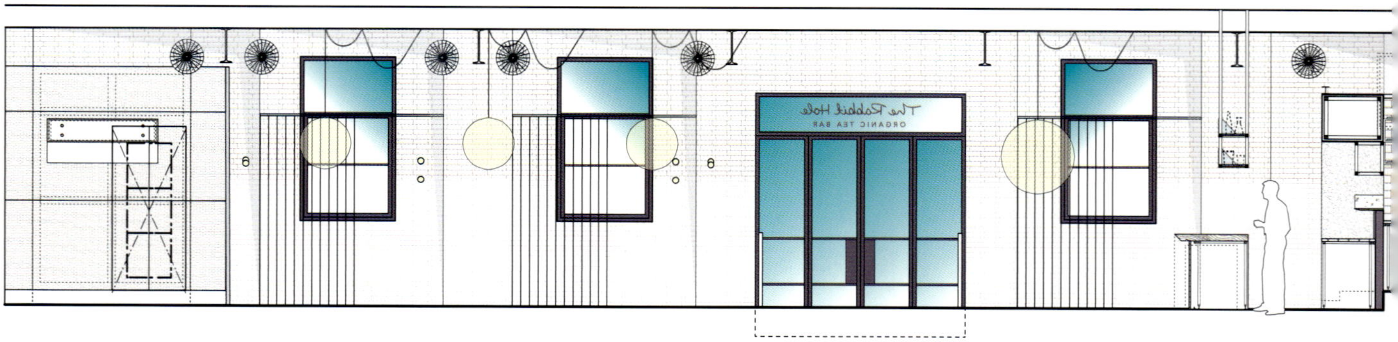

立面图
Elevation Plan

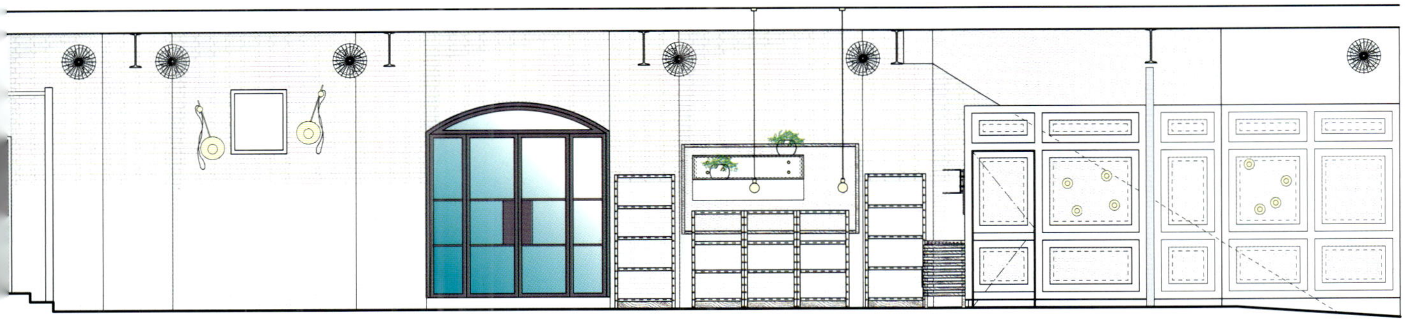

ChiChi 4U 酒吧

ChiChi 4U Bar

项目地点：波兰，波兹南

项目面积：70m^2

设计公司：mode:lina ™

设计师：Paweł Garus, Jerzy Woźniak, Kinga Kin, Magdalena Skop

摄影：Marcin Ratajczak

Location: Poznan, Poland

Area: 70m^2

Design Studio: mode:lina ™

Designer: Paweł Garus, Jerzy Woźniak, Kinga Kin, Magdalena Skop

Photographer: Marcin Ratajczak

波兹南市的威尔达区有一个属于汉堡包和比利时薯条爱好者的聚集地——ChiChi 4U 酒吧。黑色波纹钢覆盖了大部分墙壁，原始粗犷的工业感觉被木材和绿色植物中和。除了纯粹的装饰价值之外，这些钢表面可洗涤且十分耐用，钢板也让旧墙和电气布局得以快速便捷地调整。由黑色钢丝网制成的悬吊天花板露出了银色的工业管道，增强了原始工业的氛围。

District of Wilda in city of Poznan has a new venue for burger and Belgian fries lovers – ChiChi 4U. The black corrugate steel covered most of the walls. Raw feeling has been been mellowed by wooden elements and plants. The steel except for its purely decorative value brings in some extra functionality as a durable and washable surface. In addition steel sheets allowed a quick adaptation of the old walls and electrical layout. Suspended ceiling made of black steel mesh reveals silver technical ducts expand the raw and industrial atmosphere of the place.

WC
Śniadanie
frankfurterki
+ sałatka
+ grzanki
8,99
Kanapki
ser żółty
szynka
twaróg
6,99
Wi Fi

ChiChi 4U
CHIPS & CHICKEN

为了给这个坚固硬朗的内部空间增加一些家常和温暖的感觉，地面局部铺设了类似于地毯图案的花砖。黑色内饰带有一些红色的点缀，突出了品牌的视觉形象。

To give this rugged interior some homely and warm touch the foor has been covered with carpetshaped patterned tiles. Black interior with some red splashes refers to brand's visual identity.

热带雨林餐厅

Bar Botanique Cafe Tropique

项目地点：荷兰，阿姆斯特丹

项目面积：350m²

设计公司：Modijefsky 工作室

设计师：Esther Stam, Natalia Nikolopoulou, Kristina Petrauskaite, Moene van Werven, Carlotta Bassi

摄影：Maarten Willemstein

Location: Amsterdam, The Netherlands

Area: 350m²

Design Studio: Studio Modijefsky

Designer: Esther Stam, Natalia Nikolopoulou, Kristina Petrauskaite, Moene van Werven, Carlotta Bassi

Photographer: Maarten Willemstein

来自荷兰的 Modijefsky 工作室让一家阿姆斯特丹东部的餐厅变成了热带雨林。双层高的玻璃窗覆盖了整个转角处的外墙，通透是空间设计的亮点。循着玻璃窗流泻进来的日光，通过天花上悬挂的反光镜片的折射，营造出一种雨林中光线透过树枝投射下来的效果；街道对面驶过的火车，接连快速闪过的画面，形成闪烁的反光，这是空间色彩、材质和镜面反射的绚丽风景。一抹亮蓝色的休闲椅出现在棕色大理石吧台的前面，提升了色彩带来的吸引力。在顾客定制咖啡的角落里，明黄色的咖啡机闪闪发光，给空间增添了另一种色彩。

A tropical addition to the restaurant scene in Amsterdam East, Bar Botanique, designed by Studio Modijefsky. Double high large windows cover the entire façade of the corner building and spill in plenty of daylight, creating a direct link between the interior and its surroundings. The windows are a focal point in the design and concept of the interior, mirrors hang from the ceiling reflecting light and the scenery. A train, running by opposite the café, creates a flickering reflection which is captured in the bar back wall with the diverse use of colour, fabric and mirror strokes. A stroke of bright blue, the lounge chairs, appear in the interior in front of the bar with the brown marble top, giving it yet another colour boost. A bright yellow coffee machine shines in the custom made copper coffee corner, adding another touch of colour to the space.

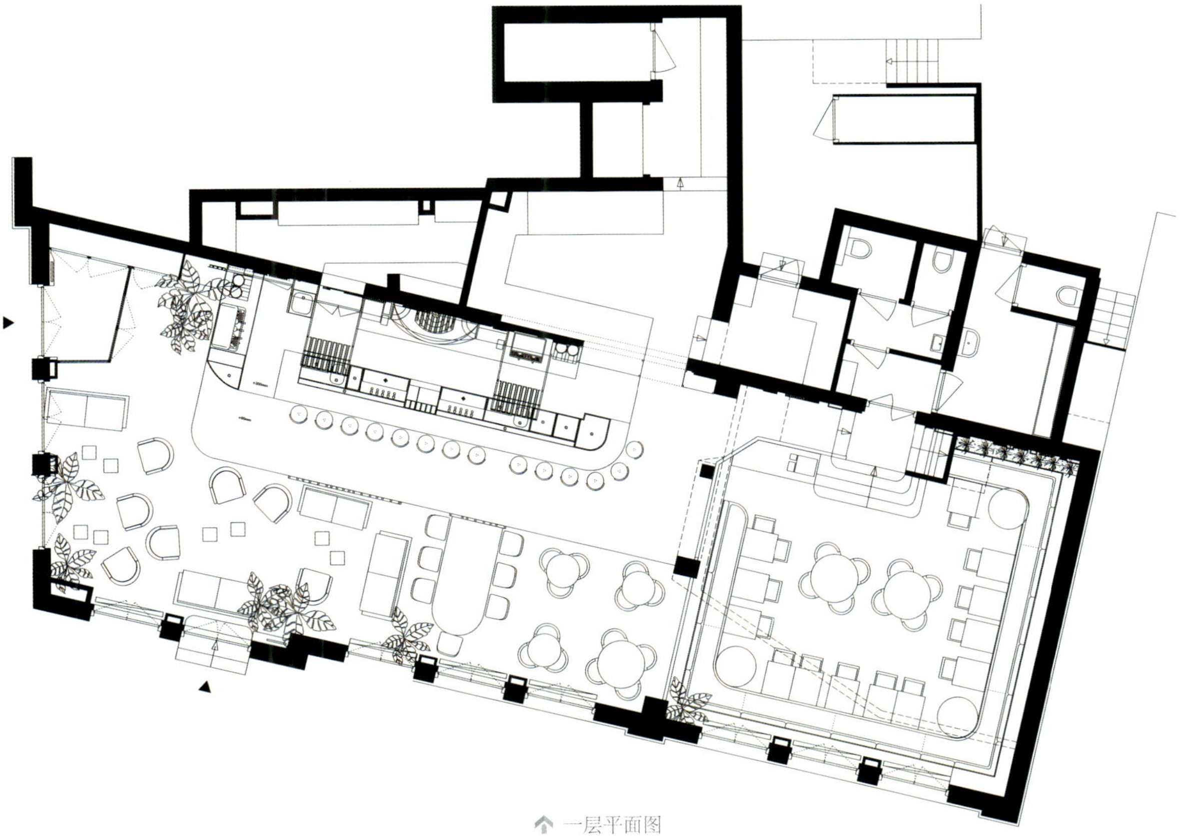

一层平面图

First Floor Plan

整个空间被栏杆以俏皮的方式来重新划分和定义。在咖啡馆旧栏杆的启发下，新的栏杆与天花板相连，并在酒吧和休息区之间形成一条清晰的界限。温暖的橡木地板覆盖了大部分区域，而明亮的绿色混凝土地板自然衔接着内部的不同区域。

The space has been divided and defined using railings in a playful way. Inspired by the old railings in the café, the new railings reach up to the high ceiling and create a clear path between the bar and lounge area. Warm oak plinths cover the floor while a bright green concrete floor naturally connects the different areas of the interior.

位于夹层和地下室的餐厅与其他室内空间分离开来。温暖的粉红色墙面，植物装饰的菜单和绿色的天鹅绒长椅赋予地下餐厅独特的个性。壁灯是用简单的圆形玻璃制成的，在平滑的玻璃上雕刻出同心圆肌理，透过黄色的灯光，在墙壁上投下一直延伸到天花板的令人愉悦的圆形阴影。

Located on the entresol and souterrain the dining area is separated from the rest of the interior. Warm pink walls, a flower press menu-board and green velvet bench rolls give the souterrain a special character of its own. Custom light fixtures which are made using simple glass plates with engraved circles on them create pleasant circular shadows along the wall up to the ceiling.

植物在室内是一个不可忽视的存在。棕榈树、喜林芋属和蕨类植物在创造一个不断变化的空间。在灯光的照射下，植物在天花板和墙壁上产生了戏剧性的效果，随着太阳下山，阴影变得更加生动。

Plants have a strong presence in the interior. Palm trees, philodendron and ferns create an ever changing interior. Highlighted with lamps, the plants create dramatic effects on the ceiling and walls, as the sun goes down the shadows become more present.

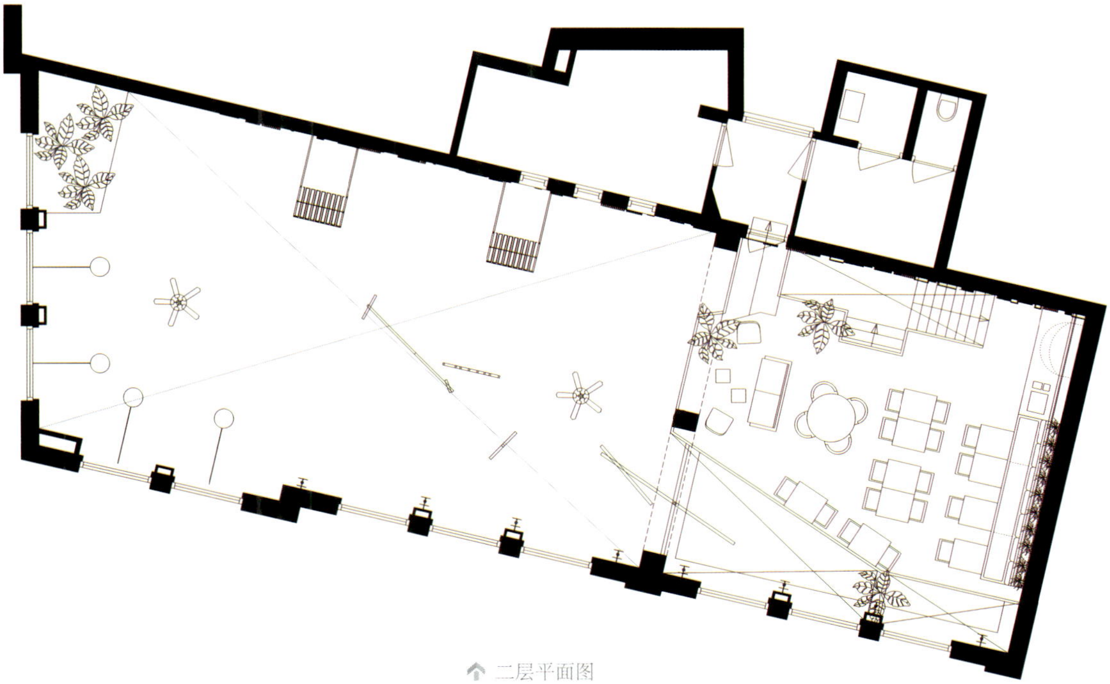

二层平面图
Second Floor Plan

Kanarie 俱乐部

Kanarie Club

项目地点：荷兰，阿姆斯特丹

项目面积：699m²

设计公司：Modijefsky 工作室

设计师：Esther Stam, Natalia Nikolopoulou, Kristina Petrauskaite, Moene van Werven, Kleoniki Fotiadou, Zahra Rajaei

摄影：Maarten Willemstein

Location: Amsterdam, The Netherlands

Area: 699m²

Design Studio: Studio Modijefsky

Designer: Esther Stam, Natalia Nikolopoulou, Kristina Petrauskaite, Moene van Werven, Kleoniki Fotiadou, Zahra Rajaei

Photographer: Maarten Willemstein

Kanarie 俱乐部位于阿姆斯特丹西部的前电车站 De Hallen 中，俱乐部以新的设计向旧的电车站致敬。Modijefsky 工作室深入挖掘建筑历史，创造了一个带有强烈历史感的设计。作为一个电车站，这里有许多为电车服务的设施，例如电力车间、油漆车间、木材车间和金属焊接车间等。这些必需的技术、材料和工作间都影响着新空间的材料和颜色选择，而且也是新的餐厅空间区域划分的参照。

Set in De Hallen, a former tram depot in Amsterdam west, Kanarie Club pays tribute to the old trams with its renewed design. Digging deep into the history of the building, Studio Modijefsky has created a design with strong references to the past. As a tram depot there were several activities taking place in the space giving service to the run down trams, such as electricity, paint, wood and metal work. These skills and the tools and work spaces they require, inspires the material and colour palette of the new interior and are means to divide the new restaurant space into different zones.

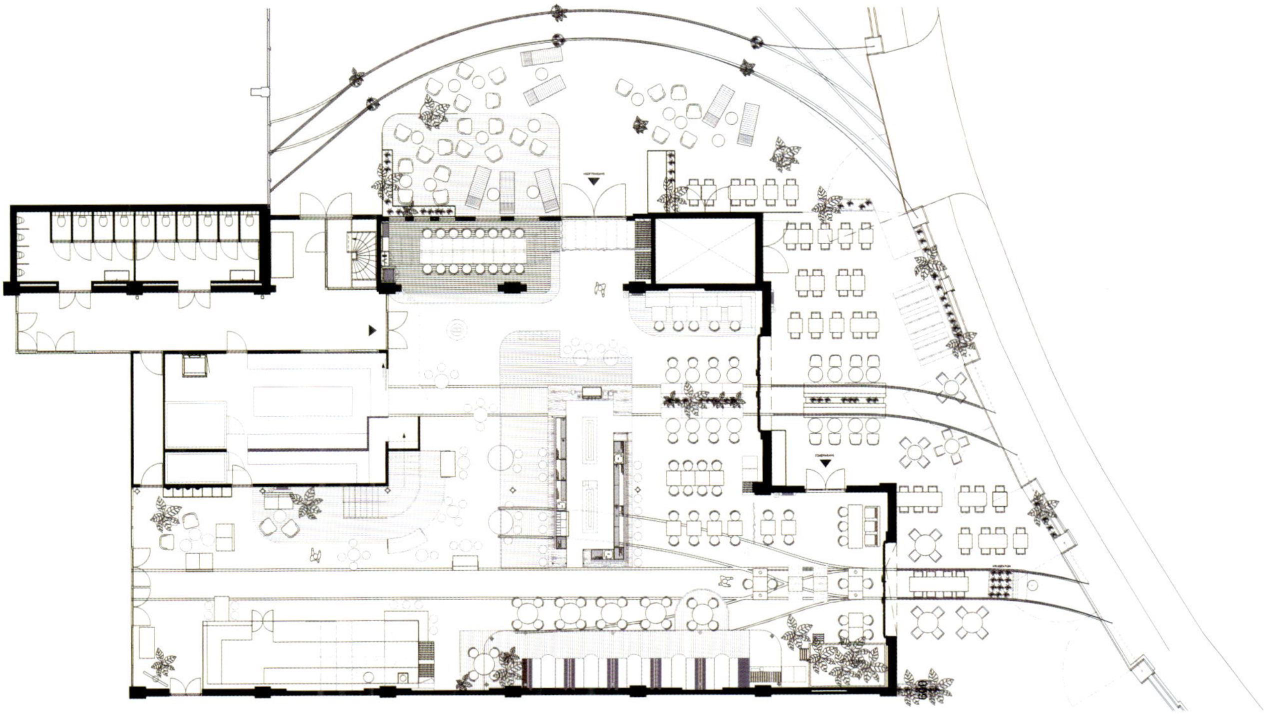

一层平面图

First Floor Plan

KANARIE

PAINT
MIND YOUR STEP

轨道引导电车进入车站并接受检查，俱乐部设计保留这些电车轨道，并应用到新的设计中，创造出挺拔的线条，它们爬上吧台并形成一个布满裸露灯泡的拱形门廊，这个由铁轨组成的巨大拱门打破了空间，并照亮了吧台区域，增强了吧台的存在感。

What remains today are tram rails that led the trams into the space to undergo treatment. The tram rails have been respected in the new interior and celebrated in the design while creating strong strokes, climbing up the bar and forming an arch with integrated lights which frame it. This great arch, formed by the rails, breaks the space and shines onto the bar, highlighting its presence.

电车标牌和语言已被用作餐厅的标签，而定制的家具则是参照老式电车座椅的复古设计。在部分有轨电车轨道上，一些座椅安装了轮子，似乎可以随时滑下铁轨，给设计带来一丝玩味。带有彩色笔触的工业窗帘，在空间中形成明确的分隔。

Tram signage and language has been used for the restaurant signing, while custom made furniture resembles the vintage design of the old electric tram seats. Positioned on the remaining tram rails, some seats stand on small wheels and seem as if they can slide down the rails any moment, giving the design a touch of playfulness. Industrial curtains with coloured strokes, create a clear separation in the space and form the vestibule for those entering the restaurant from the terrace.

俱乐部在白天主要作为工作场所使用，工作区有一个带有内置 usb 端口和插头的大型办公桌，办公桌上方的照明设计继续沿用电车轨道元素。受到油漆车间的启发，这个空间完全被绿色所覆盖，呈现出一个淡绿色的宁静空间。为了方便顾客寄存手机、手提电脑等贵重物品，俱乐部内还专设了一个带充电功能的储物柜，你可以将正在充电的手机或者手提电脑锁在储物柜中。

Designed to be a work place during the day as much, A large group table with built in usb ports and plugs has been placed in the work zone at Kanarie Club. The form and repetition in tram rails has been used as a strong reference for the design of lights above the group table. Inspired by paint workshops this space is completely covered in green and appears as a splash of light green paint in the space. special facilities have been thought of, including a charging point with lockers where you can lock up a phone or a laptop while it charges.

OCK&I2
HARGE3
TATION
5 6 7 8 9 10
2 14 15 16
ONLY
BOTTOMS UP
AT THE POOL

在厨房上方设计了一个带有休息区的错层式池畔酒吧。这里有游泳圈造型的舒适靠垫和无处不在的蓝色元素，是放松身心并享用鸡尾酒的理想场所。逼真的游泳池由蓝色橡胶制成，搭配水蓝色瓷砖，并配有典型的游泳池栏杆和带水深刻度的水波纹金属镜面电梯门。

A split level is created above the kitchen to form an entresol with a lounge area called the Pool Bar. With round comfy cushions and a splash of blue everywhere, it's the perfect place to unwind and enjoy a cocktail. Made out of blue rubber with a stroke of matching tiles, the pool is complimented with a typical pool railing and a wavy mirror element on the bar lift.

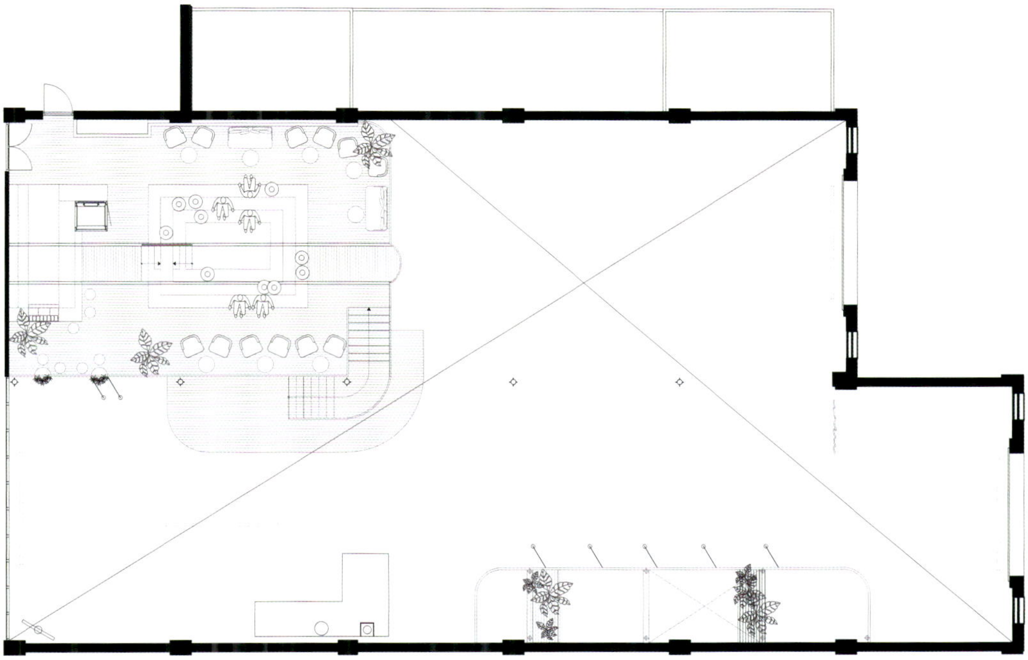

池畔酒吧平面图
Pool Bar Plan

烤肉餐厅

The Roast Room

项目地点：荷兰，阿姆斯特丹

项目面积：603m²

设计公司：Modijefsky 工作室

设计师：Esther Stam, Natalia Nikolopoulou, Kristina Petrauskaite, Kasia Nowak

摄影：Maarten Willemstein

Location: Amsterdam, The Netherlands

Area: 603m²

Design Studio: Studio Modijefsky

Designer: Esther Stam, Natalia Nikolopoulou, Kristina Petrauskaite, Kasia Nowak

Photographer: Maarten Willemstein

一楼是一间小酒馆风格的餐厅。坐在大型砧板形状的餐桌旁，顾客可以看到厨房和吧台，一侧吧台上方悬挂着一个大型的铜制啤酒罐，另一侧吧台上的玻璃展柜和金属架上放置了店内贩卖的生鲜肉品。半透明的工作间一侧有一个巨大的转轮，转动轮子可以将玻璃窗开启，顾客可以直接看到厨师工作的场景。

On the first floor of The Roast Room you will find a brasserie-style restaurant. Sitting at large cutting boards you have a view of the kitchen and a bar with large copper beer tanks hanging above it. In the marble bar, display windows showcase the goods of the in-house butcher’s shop. By turning a giant wheel, the butcher opens his shop window so guests can see him cutting and boning the meat.

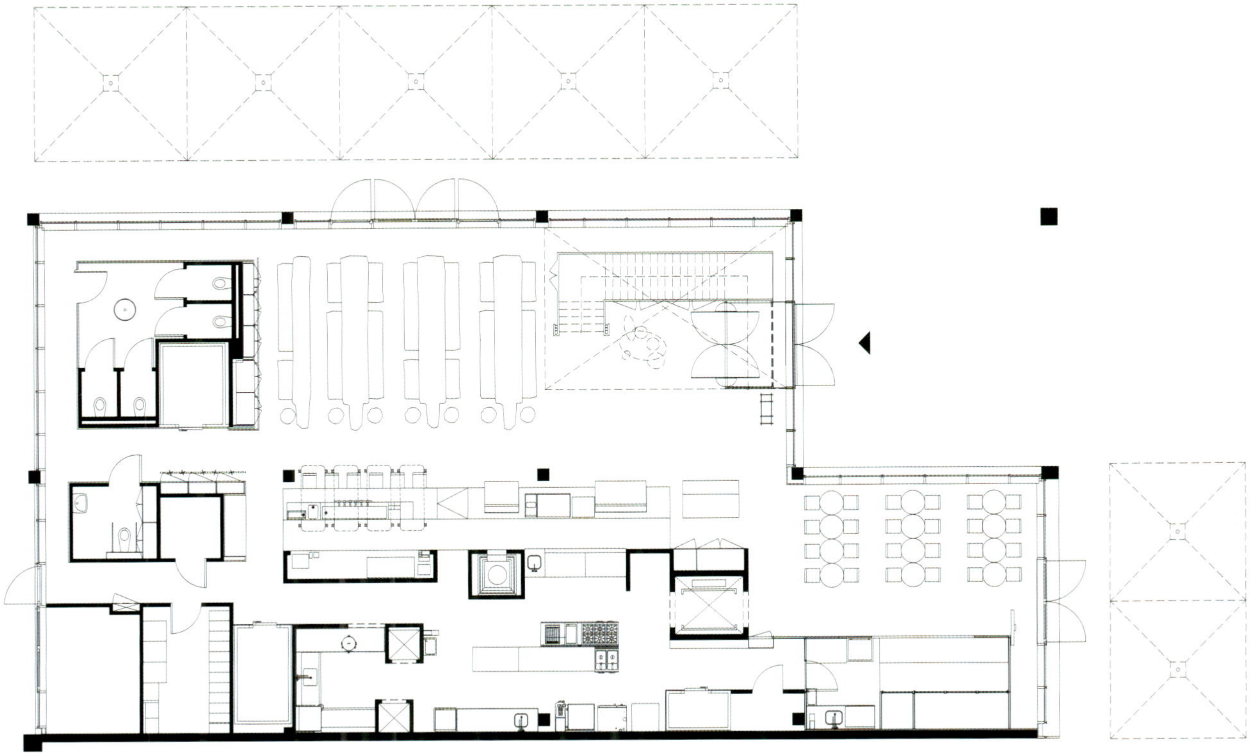

一层平面图
First Floor Plan

TOILET

BUTCHER

Ernst-Jan
T.Heuft
J.Krouwels
M & B

顺着如同砧板般材质肌理的楼梯通往二楼的烤肉餐厅。温润的奶油色吧台和带有金属质感的深色开放式厨房沿着空间内侧一字排开，空中悬挂的一盏盏黄铜色的吊灯，为顾客创造出一个个精致的小空间。而整个餐厅中最引人瞩目的，则是悬挂在楼梯井处，联系上下空间的姜块飞船吊灯，与其他的铜制品一样，这个巨型吊灯也是由 Modijefsky 工作室为餐厅量身定制的。

A stairway resembling a chopping block leads to the Rotisserie on the first floor. Here, a creamy white bar, a pitch-black kitchen and a sky of brass light fixtures make up an intimate setting for some fine dining. The centrepiece is a light sculpture, Ginger Blimp, which connects the two floors. Studio Modijefsky made this monumental chandelier, just like the other bronze works, especially for the restaurant.

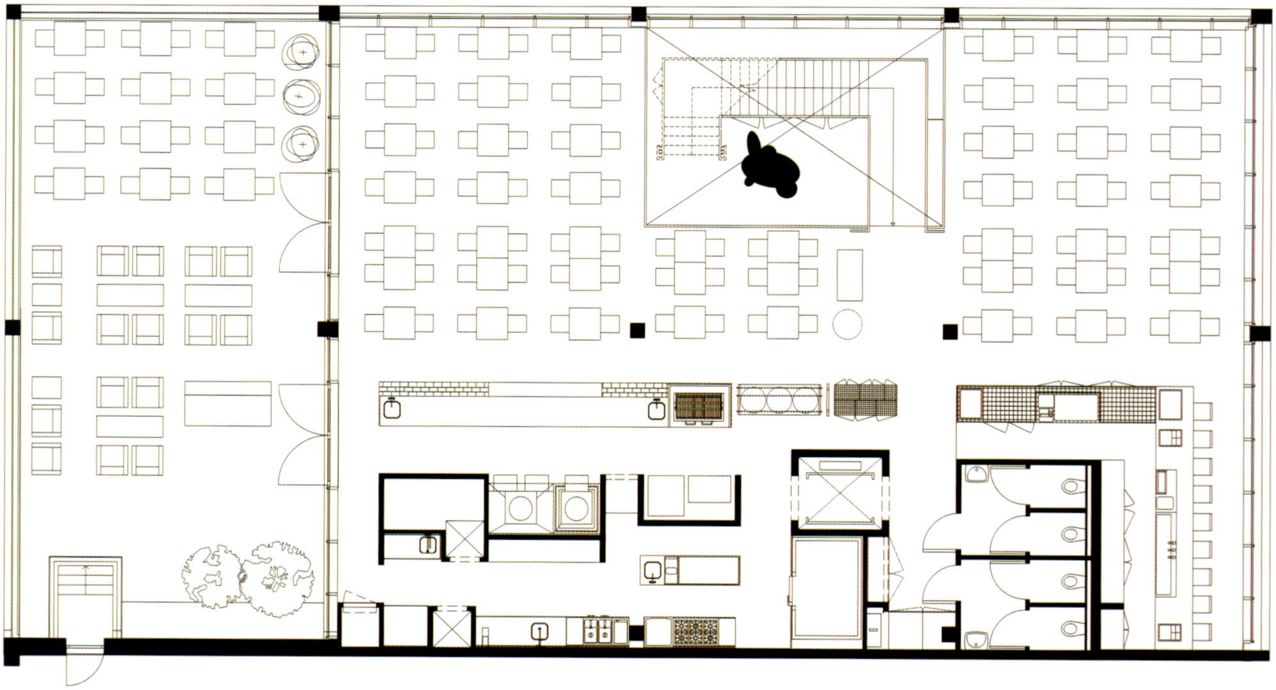

二层平面图

Second Floor Plan

TOILET

TOILET

透明度和耐用性是餐厅的主要特点，这在餐厅的设计中明显可见。设计公司创建了一个开放、粗犷且经久耐用的室内空间。

Transparency and durability are the main ingredients for The Roast Room and this is made visible in the restaurant's design. Studio Modijefsky created an interior with an open, robust and timeless character.

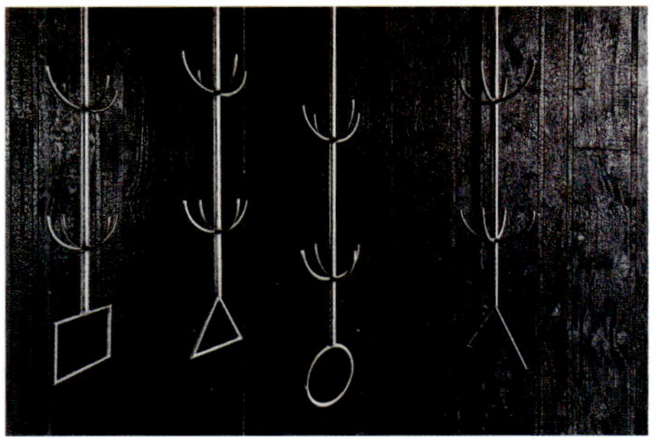

A 空间——柏林阁楼

A Space – Lofts in Berlin Mitte

项目地点：德国，柏林

项目面积：400m^2

设计公司：plajer & franz studio gbr

摄影：Christian Rudat

Location: Berlin, Germany

Area: 400m^2

Design Studio: plajer & franz studio gbr

Photographer: Christian Rudat

本案地处德国柏林市中心米特区一座废弃的剧院工业建筑内，现在已被全部重新设计，作为 natulis 集团的临时售楼展厅。斑驳粗糙的旧工业建筑与全新的高品质材料对比，产生强烈的视觉效果。原来的天花板被拆除，8m 高的空间就像原始剧院的大荧幕，墙壁保留了裸露的砖木结构，甚至包括过去几十年的维修痕迹。

A former movie theatre, in a previous industrial building in Berlin Mitte has been fully redesigned and given a new image. The location give home to a temporary showroom for the natulis group ag. A fascinating symbiosis between industrial attributes and highquality materials is responsible for the distinctive character trait. To achieve this the original space has been gutted, the inserted ceiling removed and a room with almost 8 meters ceiling height-corresponding the original state as a theatre screen - created. The walls have been left in its' primar, brick-lined condition showing the exposed brick wood including traces of repair works from the last decades.

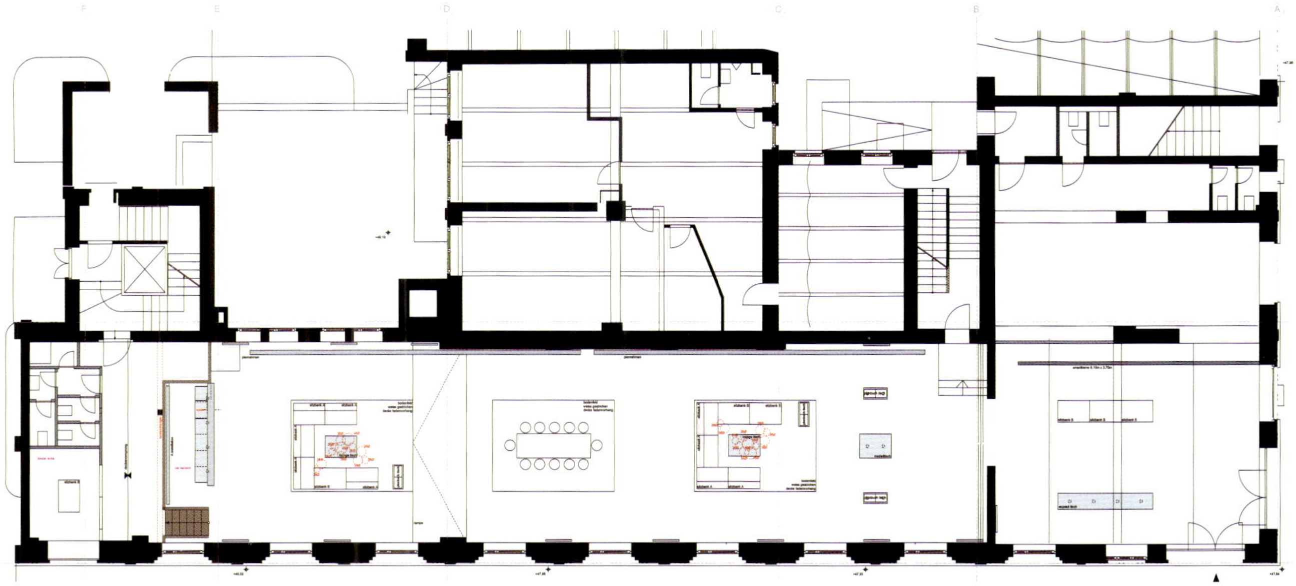

平面图
Floor Plan

沙发由刷深色油漆的木托架制成，并配有高品质的沙发坐垫；所有的桌子由黑钢制成；长达 5m 的线帘悬挂在洽谈区的上方，划分出了不同的空间；超大的“A”形灯泡装置挂在接待台正上方，让我们想起 20 世纪 20 年代电影院的照明广告，从而揭示了这个空间的原始功能。

Seating furniture are made of dark oiled standard industrial palettes, yet covered with a premium-quality upholstery. all tables are made of black steel. Noble, up to five meters long thread curtains hang above the counseling islands and serve as partitions structuring the space. A light bulb installation in the shape of an "A" reminds us of illuminated advertising known from cinemas of the 1920s and in thus reveals the original function of the space.

展厅在 arkona 庭院（A 空间的位置）提供展示和出售阁楼的功能。所以设计必须符合这一功能目的，提供宽敞的演讲区域以及与客户进行沟通的休息区。该地区的城市地图在入口处迎接客户，而墙上的建筑物和平面图的模型使人们对出售的公寓有了直观感受。在宽敞的开放空间的尽头，是一个小酒吧，供应饮料和小吃。

The showroom serves the presentation and sale of lofts in the arkona courts (the location of A space). The design had to correspond this functional purposes and offer spacious presentation opportunities as well as lounge areas for client communication. A city map of the area greets the customers at the entrance while a model of the building and floor plans on the walls are giving a visual idea of the apartments for sale. At the end of the spacious open space is a small bar for drinks and snacks accompanying consultation.

A
NATULIS LIVING

PMSA 办公室

PMSA Office

项目地点：土耳其，伊斯坦布尔

项目面积：12000m²

设计公司：mimaristudio 工作室

设计师：Ayça AKKAYA KUL, Önder KUL

摄影：Gurkan AKAY

Location: Istanbul, Turkey

Area: 12000m²

Design Studio: mimaristudio

Designer: Ayça AKKAYA KUL, Önder KUL

Photographer: Gurkan AKAY

该项目是 PMSA 新的办公室和运营总部，其设计、起草和项目管理过程由 mimaristudio 工作室执行。新的办公室位于伊斯坦布尔的 Kağıthane Cendere 路，这条路是城市商业发展的中轴线。办公总部共有 3 层楼，面积达 12000m²，其中 4000m² 是办公室。

The PMSA office and business center was designed by mimaristudio, so as the drafting and project management. The new office headquarters is located on Kağıthane Cendere Road in Istanbul, which is a developing business and commerce axis of the city, has 12000m² of area on 3 stories, 4000m² of which are offices.

考虑到如此庞大的空间和体量，工作人员被分成两个大型特色工作组来设计，每个工作组分别被赋予独特的个性化外形。这些特色造型的设计分别与团队的物理环境、专业和技术需求相对应，满足团队或个人的工作需求。设计采用在大空间中创建小空间的方法，在工作区创建了数个开放的子空间。这样的设计手法避免了开放式空间中最常见的枯燥的展位式排列，不仅丰富了整体的空间感，还为每个团队塑造了各自的个性化特色。

Considering the largeness of the space both in area and the volume, team shells to combine two work groups are designed so that each work group can have their own identity. These shells which respond to physical of the staff, are designed with a space inside a space approach creating open sub-spaces inside the main work area. With this approach showroom-like effect is avoided and a different identify for each team is provided.

办公区的其他功能都围绕着中央核心区而设。此外，还设有一间大型开放式多功能厅，以满足团队的多种功能需求，并与旁边的活动区相连，形成大型会议区或派对区。另外位于多功能厅的旁边的三个不同颜色的球形“游戏壳”，是设计团队专门为本案定制设计的。

Other units in the office are placed around this core. In addition to this, an open multipurpose hall is proposed for the new place. This hall, which is considered for varying types of usages, is linked to social spaces. The 'game shells' which were specifically designed by mimaristudio are placed on the area that is next to this space.

地面材质的设计既有变化，又有内部的关联性。犹如街道般开敞的走道，地面装饰延续了会议室和休闲区的地面材质，用 5mm 厚的 PVC 方块地板满铺，既易于清洁又有降噪的作用。材质的色彩和图形根据不同的空间设计理念而有所变化。

The space between the shells that cover working clusters are composed like streets, the material selection for these spaces are arranged to create a sense of continuity between other meeting and social spaces. 5 mm modular PVC tiles are chosen for the all floor finishing, expect meeting and interview rooms, for its sanitary and acoustic advantages. Their color and texture vary to go along with the interior design concept of each space.

照明设计标准是按照 mimaristudio 工作室以前的成功经验来设定的。所有的照明均是 LED 光源，工作空间选用点光源配合线性光源照明。在开放空间，则选用不同规格和形式的特色灯具装饰，其中一些灯具还具有声学方面的优势。在总预算范围内，设计师重新对家具进行了一些添加和更改。除了购买成品外，咨询处、会议室、茶水间和复印处的物品都是定制的。它们的颜色和质感随着每个空间的设计理念变化。

Lighting design criteria are set by mimaristudio team as it was in the previous mimaristudio projects. LED products are used for all of the spaces in the project, linear or spot projections products are chosen for the working spaces, while in open spaces more specialized products with varying sizes and forms some of them having acoustic qualities are preferred. Staying within the total budget, were reused with some additions and alterations while new suitable products are selected. Besides purchased products, the objects that were designed for information desk, meeting rooms, drinking and copying spots are all custom made. Their color and texture vary to go along with the interior design concept of each space.

>system room
IS

Opera 波兰分部办公室

Opera Software Poland Office

项目地点：波兰，弗罗茨瓦夫

项目面积：4000m^2

设计公司：mode:lina ™

设计师：Paweł Garus, Jerzy Woźniak, Kinga Kin, Małgorzata Szymańska

摄影：Marcin Ratajczak, Maciej Dąbrowski

Location: Wroclaw, Poland

Area: 4000m^2

Design Studio: mode:lina ™

Designer: Paweł Garus, Jerzy Woźniak, Kinga Kin, Małgorzata Szymańska

Photographer: Marcin Ratajczak, Maciej Dąbrowski

Hieronimus 商业中心（也被称为玻璃房子）是居住在弗罗茨瓦夫的人们心目中极其重要的建筑。Opera 波兰办公室就在其附近。城市风貌在办公室设计过程中给了设计师许多灵感。鉴于员工的风格偏好和城市氛围，建筑师最终确定其室内空间为工业风格。弗罗茨瓦夫作为“百桥之城”的形象将成为该项目的主题，无论是拱门、壁炉的形式还是色彩的选择都取自城市中最受欢迎的空间。

Business centre Hieronimus (the so-called Glass House) is an extremely important building for people, who live in Wroclaw. As a result, the city architecture has inspired the designers during the design process of the company's interior. Given the stylistic preferences of employees and the atmosphere of the city- the final result has been preserved in the industrial. Thus, the picture of Wroclaw as "the city of one hundred bridges" has became the main theme of the project. Constructions in the shape of arch, grates and colours directly referring to the most popular spices are dominating inside the building.

展示空间

Representative Space

弗罗茨瓦夫火车总站
Wroclaw Glowny Railway Station

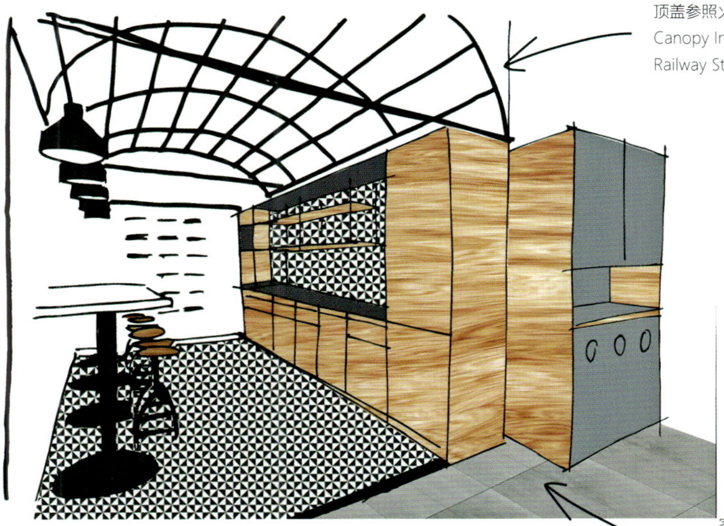

顶盖参照火车总站大厅的样式
Canopy Inspired by the Shape of Wroclaw Railway Station Main Hall

家具摆放模仿火车站的形式
Material Set Like in Wroclaw Railway Sation

↑ 就餐区
Dinning Area

拱顶模仿购物厅的大拱架
Arches of the Historic Market Hall

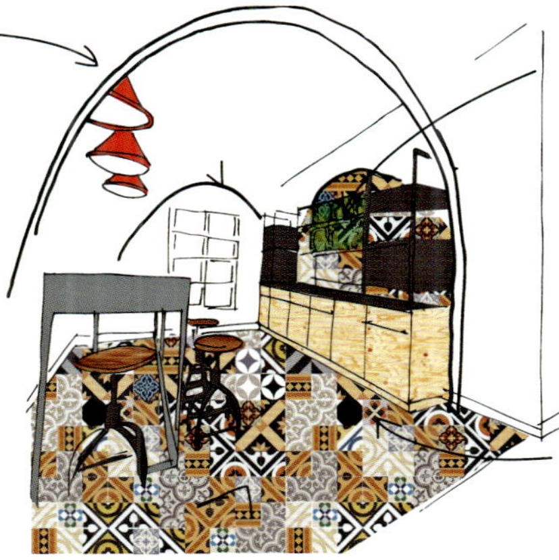

经典的水果和蔬菜储物盒
Typical Fruit and Vegetable Boxes

多彩的地砖像是购物厅的摊位
Colorful Tiles Like Market Stalls

弗洛瓦茨夫购物厅
Wroclaw Market Hall

厨房
Kitchen

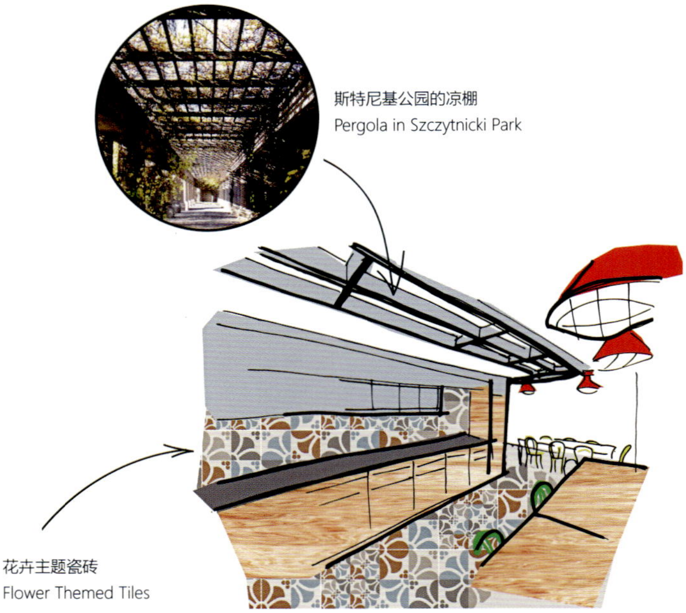
斯特尼基公园的凉棚
Pergola in Szczytnicki Park
花卉主题瓷砖
Flower Themed Tiles

↑ 厨房
Kitchen

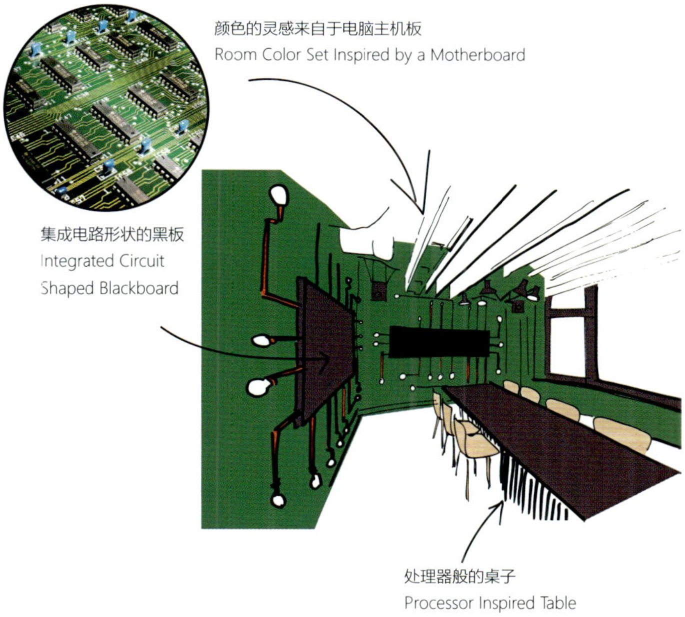

会议室
Meeting Room

墙壁的装饰模仿电脑上的冷却系统零件
Wall Decors Inspired by Computer Cooling Units

会议室
Meeting Room

会议室的设计非常具象，直接使用了大量的电脑元件和 IT 元素：比如集成电路、散热器等装置。设计师使用了将近一千个旧软盘和二百多个键盘，最终完成了这些装置的制作。

With the concrete, meeting rooms are directly associated with computers and IT branch: there is an integrated circuit, fans and decors, which are made out of nearly one thousand old foppy disks and over two hundred keyboards!

走廊的设计来源于桥梁的细节
Corridor Finishing Inspired
by Bridge Details
隔音通话间
Soundproof
Telephone Booth

走廊
Hallway

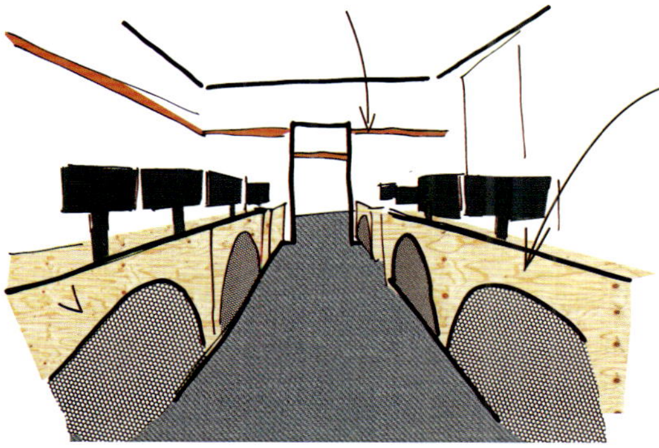

装饰源于桥梁的细节
Subtle Decor Refers to Bridge Details

桌子主体的形状参考弗罗茨瓦夫的桥上钢缆
Desk Enclosures in the Shape of Conceal Cabling on the Bridge in Wroclaw .

办公室
Office

Gazeta.ru 新闻社办公室

Gazeta.ru News Agency Office

项目地点： 俄罗斯，莫斯科

项目面积：600m^2

设计公司：Nefa Architects (Nefaresearch)

建筑师：Maria Nasonova, Marganta Kornienko

设计师：Elena Potemkina, Dmitry Ovcharov, Maria Nasonova, Margarita Kornienko

摄影：Ilya Ivanov

撰文：Dmitry Ovcharov, Elena Potemkina

Location: Moscow, Russia

Area: 600m^2

Design Studio: Nefa Architects (Nefaresearch)

Architects: Maria Nasonova, Marganta Kornienko

Designer: Elena Potemkina, Dmitry Ovcharov, Maria Nasonova, Margarita Kornienko

Photographer: Ilya Ivanov

Authors Team: Dmitry Ovcharov, Elena Potemkina

设计师所面临的挑战是为一个新闻记者小组创造一个舒适和多功能的工作空间，其中包括不同的活动区、集中工作区，以及各种会议和社交场所。Gazeta.ru 编辑部位于一栋四层建筑中的第三、四层。办公室设施包括个人和团队的工作区、会议室、休息区、迷你咖啡室以及一个位于四楼带有厨房的餐厅。此外，员工还可以使用配备有现代视听设备的多功能厅进行会议、讲座和演示。

The challenge for the designers was to create a comfortable and multifunctional work space for a journalists' team that included areas for different activities – zones for concentrated work along with various meeting and social spaces. The editorial office of Gazeta.ru occupies the third and the fourth floors of a 4-storey building. Office accommodation comprises workspace areas for individuals and teams, meeting rooms, lounge-zones, mini coffee points as well as a kitchen-dining-room located on the 4th mansard floor. Besides, employees can use a multifunctional hall equipped with most modern audio-visual equipment for conferences, lectures, meetings and presentations.

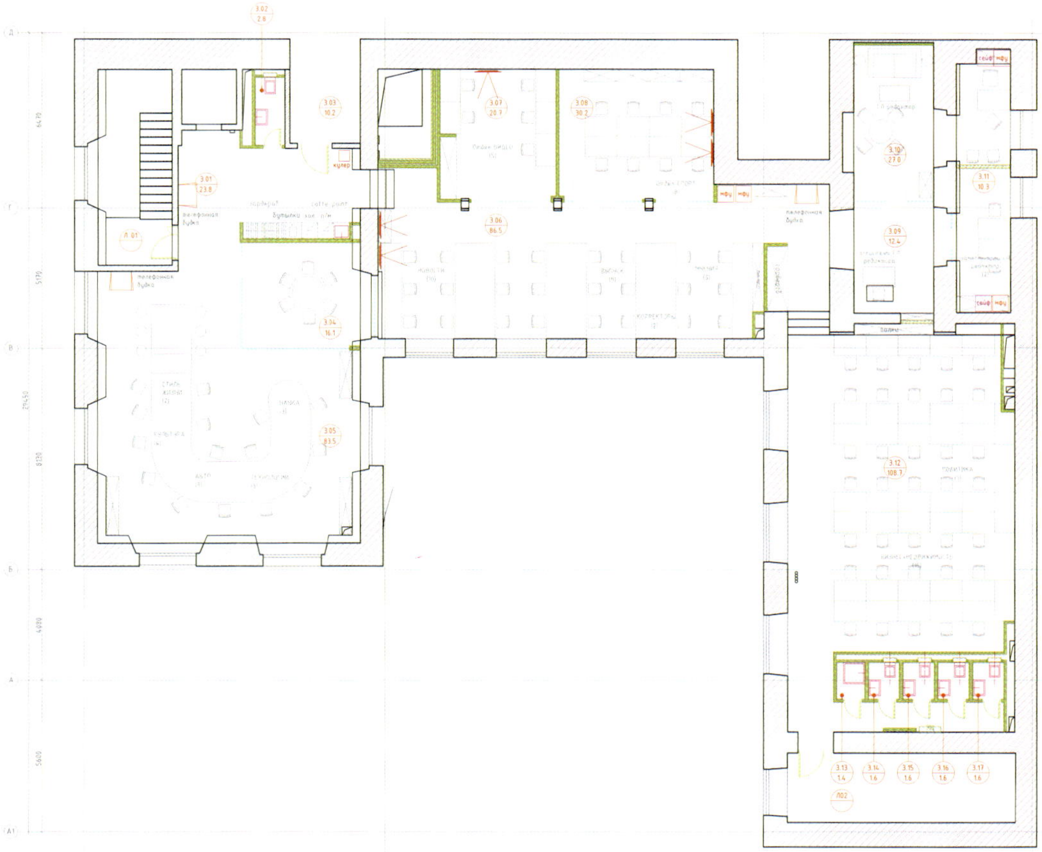

三层平面图
Third Floor Plan

GA
газета.ru

整个出版过程非常迅速，各部门之间的沟通速度极为重要。这就是工作台需要设置在同一个开放的空间的原因。总编和他的助理团队位于办公室的中心，方便与新闻部的所有部门进行交流和信息传递。

The whole publication process is very fast so the speed of communications between different departments is extremely important. That's why working desks are placed in an open space. The chief editors and his assistants' cabinets are located right in the middle of the editorial office for the convenience of communications and information transfer with all the departments of the news agency.

Gazeta.ru 作为一家门户网站涵盖了大量信息，包括政治、时尚和科技生活等。设计以白色为基调化繁为简，以弱化繁复的网络信息。以黑色衬托白色，营造简约而理性的氛围。墙面与地面上的文字图形让人联想到报纸或网站的信息内容。

The portal`s reports cover a wide rangeof news:politics, style, auto news. Architects decided to show the scope of information the agency reported usingonly one color- White. Graphic black accents on the floors and windows create an image of a newspaper page or aninformationweb-site.

KITCHEN

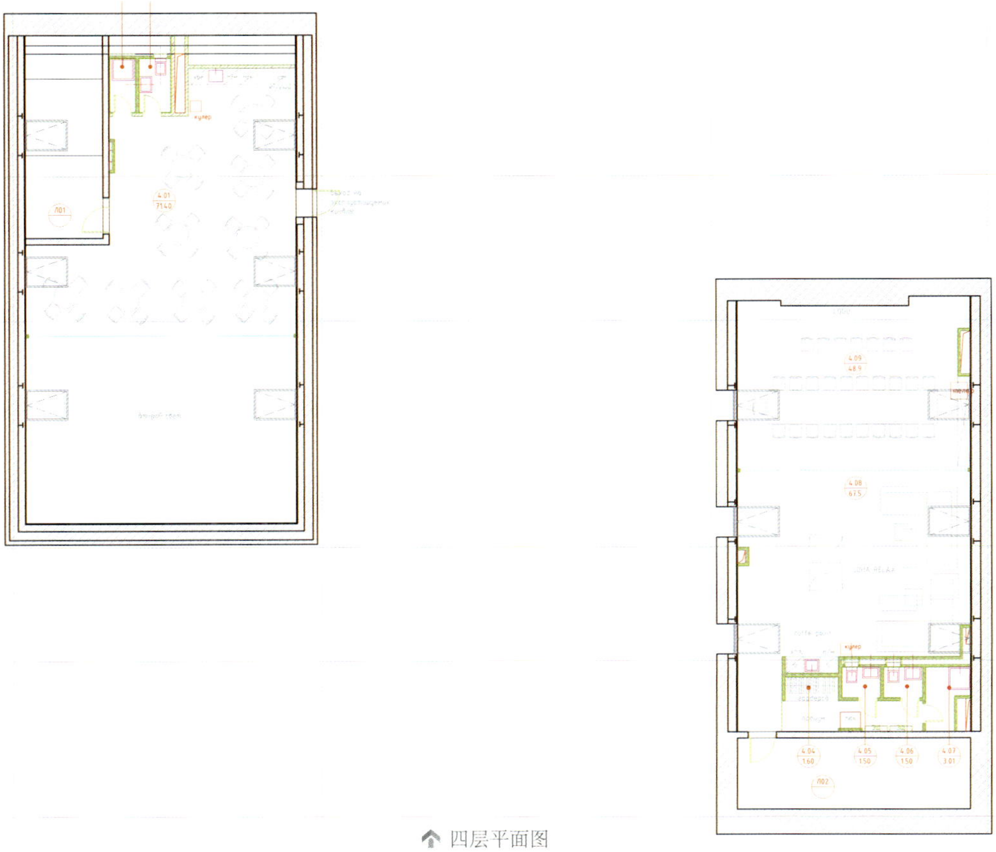

四层平面图

Fourth Floor Plan

Championat.com 体育门户网办公室

Championat.com Sports Web Portal Office

项目地点： 俄罗斯，莫斯科

项目面积：400m²

设计公司：Nefa Architects (Nefaresearch)

建筑师：Maria Nasonova, Marganta Kornienko

设计师：Elena Potemkina, Dmitry Ovcharov, Maria Nasonova, Margarita Kornienko

摄影：Ilya Ivanov

撰文：Dmitry Ovcharov, Elena Potemkina

Location: Moscow, Russia

Area: 400m²

Design Studio: Nefa Architects (Nefaresearch)

Architects: Maria Nasonova, Marganta Kornienko

Designer: Elena Potemkina, Dmitry Ovcharov, Maria Nasonova, Margarita Kornienko

Photographer: Ilya Ivanov

Authors Team: Dmitry Ovcharov, Elena Potemkina

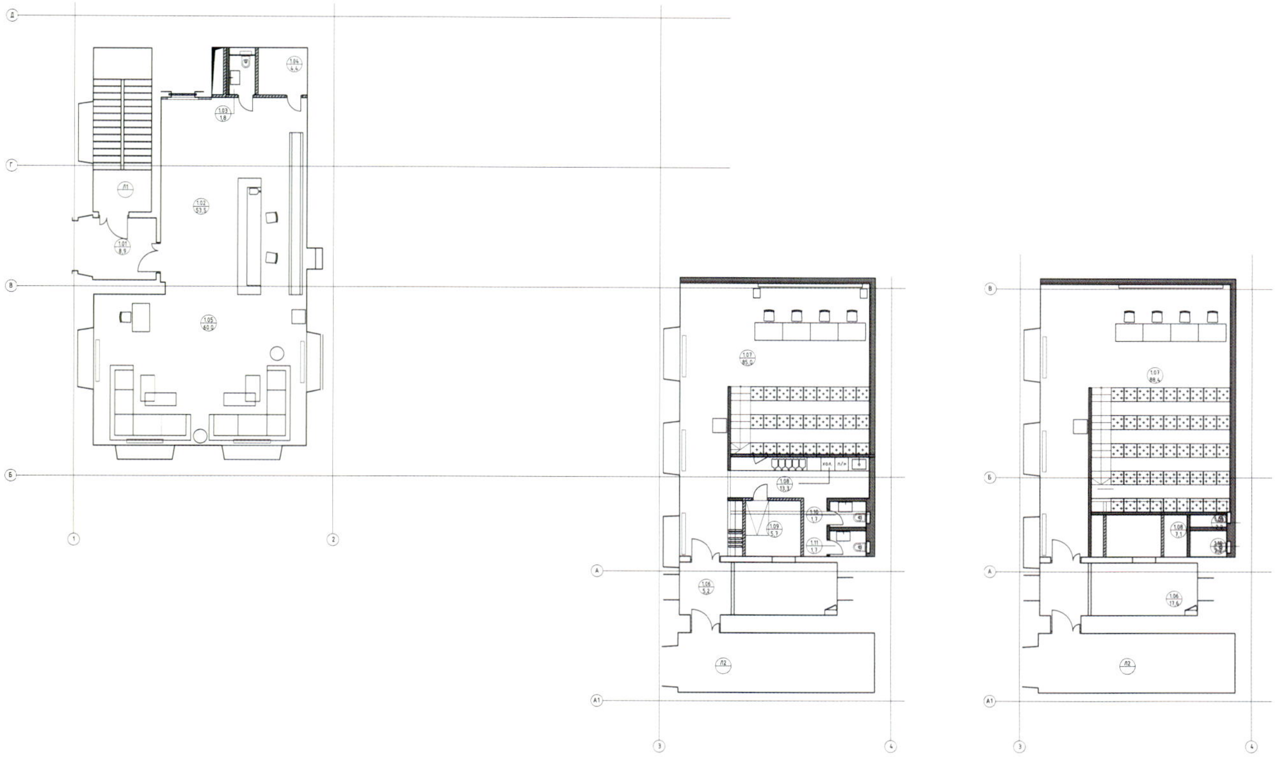

一层平面图
First Floor Plan

Championat.com 是俄罗斯著名的体育流媒体之一，主要从事体育赛事的现场直播。编辑部位于一栋四层高的 19 世纪工厂建筑的其中两层，设计的主要目标是以一个生动的主题为背景，为记者团队创造一个高效和协作的工作空间。接待处包括一个展示体育信息的显示屏和服务台，以及一处由时尚家具打造的宽敞舒适的休息区。根据客户需求，设计师设计了一个多功能厅，可用于大型会议、讲座和汇报展示，大厅配备了最先进的视听技术和设备。

Web portal Championat.com is one of Russia's leading sports streaming websites, focused on the live coverage of the sporting events.The main goal for the designers was to create an efficient and collaborative work space for a team of journalists with a vivid theme-based idea. The editorial office of Championat.com occupies two storeys of a four-storey 19th century building. Reception area includes an information desk with integrated media panels adapted to display original content about games and events and a large comfortable lounge-zone. According to the project specification the architects designed a multifunctional hall for conferences, lectures, meetings and presentations. The hall is equipped with most modern technical, audio-visual equipment.

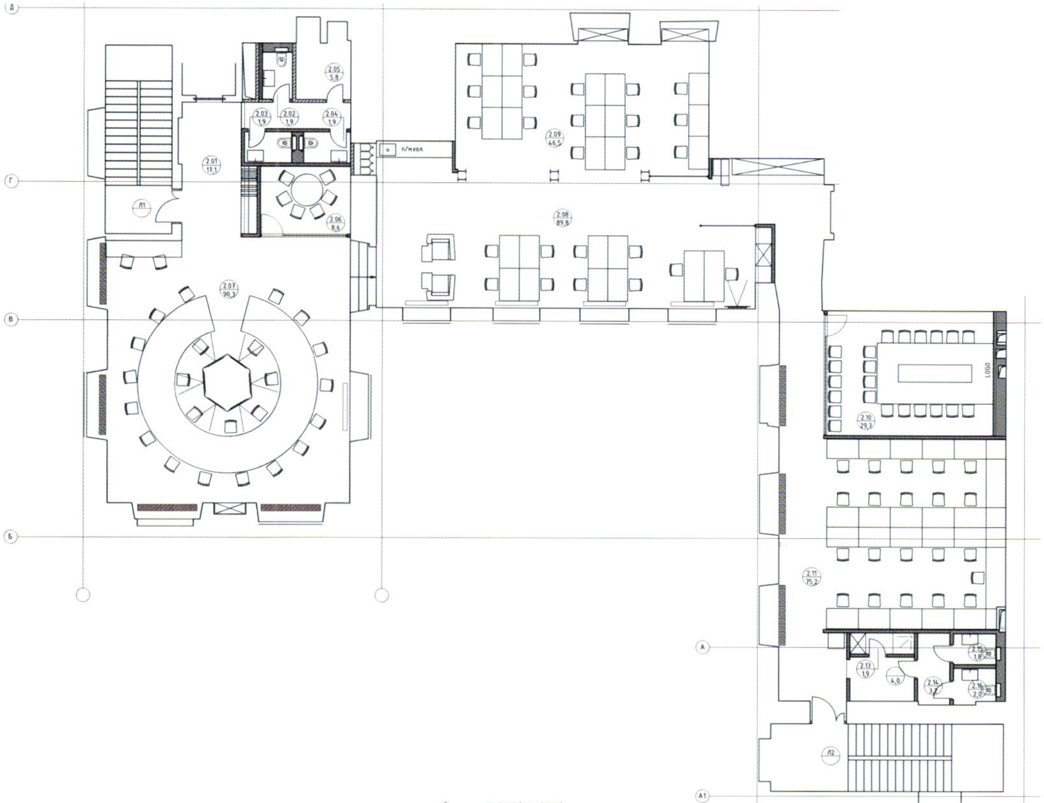

二层平面图
Second Floor Plan

工作区和会议室等主功能区位于二楼。开放式的空间设计可以让员工更好地进行沟通和交流信息。高层管理者同样位于办公室的开放式几何中心共同办公，从而使沟通过程尽可能有效和快速。中心处是一处实时监控室，这里设有一张大型的环形工作台，顶面安装有一组显示屏，这里是整个空间的核心。二楼还设有放松区和迷你咖啡室。

A monitoring roomwitha big ring-shaped table and a video installationset just below the ceilingin the middle is theideological "core" of the office. Top-managers` desks are also placed in an open-space in the geometric center of the office to make the communication process as effectiveand fast as possible.There are also a relax-zone and a mini coffee point on the 2th floor.

建筑师几乎全部使用自然材料来装饰办公空间内部，如红砖、实木、金属、玻璃等。许多办公家具是由工作室专门设计的，并在当地的车间生产。

The architects used mostly natural materials such as brick, wood, metal, glass in the interior decoration of the office space. Many pieces of office furniture were specially designed by the studio and produced in local workshops.

Habita 联合办公室

Habita Coworking Offices

项目地点：土耳其，伊斯坦布尔

项目面积：650m^2

设计公司：PAB Architects

设计师：Pınar Gökbayrak, Ali Eray, Burçin Yıldırım

摄影：Murathan Varol

Location: Istanbul, Turkey

Area: 650m^2

Design Studio: PAB Architects

Designer: Pınar Gökbayrak, Ali Eray, Burçin Yıldırım

Photographer: Murathan Varol

Habita是一个以创新为理念设计的办公空间，旨在通过协作、互动和相互学习来维持创新、创造力和生产力。空间可以持续地作为创意空间，让员工可以在专业领域、社交管理、生活方式上互相学习，共同践行“在共享环境中加速前进”的座右铭。

Habita is an office space designed with the notion of innovation, creativity and productivity that can maintain by collaboration, interaction, and learning from each other. The space aims to sustain a creative community which can learn from one another both professionally and socially in the coworking and living environment with a motto of "accelerating in a shared habitat".

1 入口大厅和接待处 Entrance lobby and reception
2 私人办公室 Private offices
3 开放办公室 Open office
4 活动工作区 Flexible working area
5 储物间 Lounge
6 活动区 Event space
7 厨房 / 吧台 Kitchen/bar
8 会议室 Meeting room
9 杂物室 Storage
10 卫生间 Toilet

平面图
Floor Plan

以这种理念设计的空间具有开放和透明的特点，布局生动而灵活，让用户可以在不同的“气氛”中工作。通过提供给用户不同的隐秘工作环境，不同的座位安排（坐在椅子上，坐在沙发上，站着，躺着），在不同的地方，满足他们对工作地点和工作方式的不同要求。这些选项可为使用者提供双重工作环境——使用私人空间，或者使用具有合作潜力的开放式办公空间。

The space designed in such a perspective has an open and transparent character enabling interactions and a flexible layout that allows its users to work in different "moods"; in different privacy settings, in different seating arrangements （on a chair, on a sofa, standing, lying）, in different locations by giving people options to choose where and how they want to work. These options enable a dual working environment for users – a private space if needed or a diversified space with the potential of collaborations if requested.

办公空间位于工业区附近，周边有金属和汽车车间环绕，这些工业特色被带入室内空间，例如刨花板、铸铁材料、绳索、聚碳酸酯板，以及以木制水果箱为容器的绿植。

The office space, with a setting in an industrial neighborhood with metal and auto workshops all around, carries the industrial character of the surrounding to indoors by natural and uncut materials like osb, iron profiles, ropes, polycarbonate plates and a number of greenery within wooden fruit boxes.

650m^2 的 Habita 办公室，有 12 间面积为 10~18m^2 的私人办公室，开放式办公空间里有 8 张办公桌，每张桌子可供 6 人使用，还有灵活的工作单元可供自由时间工作的成员使用。主要的建造动机是设计一个在工作、休息和社交之间没有严格界限的空间，空间决定工作方式。因此，开放式办公空间仅通过高差来界定区域，阶梯变成了工作、社交或休闲的中间区域。

650-squaremetered-Habita, consists of 12 private offices with a range of 10-18m^2, open offices with eight tables for six each, and flexible working units for members who wish to come irregularly. The main architectural motive was to have a design where there are no strict boundaries between working, resting and social spaces that dictates how to work, rest and socialize in terms of space use. Therefore open workspaces are defined with just a level difference where the steps are turned into intermediary zones for working, socializing or resting.

BAUHAUS

BLANCO

BLANCO

项目地点： 以色列，特拉维夫

项目面积：200m²

设计公司：ROY DAVID STUDIO

摄影：Yoav Gurin

Location: Tel Aviv, Israel

Area: 200m²

Design Studio: ROY DAVID STUDIO

Photographer: Yoav Gurin

本案最初的挑战重点在于尝试创建一个动态工作区，并连接公司内的各个部门，包括空间中心的公共工作区。设计理念是根据公司的核心价值观的基础上建立的，包括透明度和协作性，因此空间需要玻璃隔断来表现。

The initial challenge was focused on trying to create a dynamic workspace that will connect the various departments within the company. The concept of dynamicity birthed design plans that included a communal workstation in the center of the space. The design concept was developed and based upon the company's core values which include transparency and collaboration; thus the space entails in it glass partitions that embody these values.

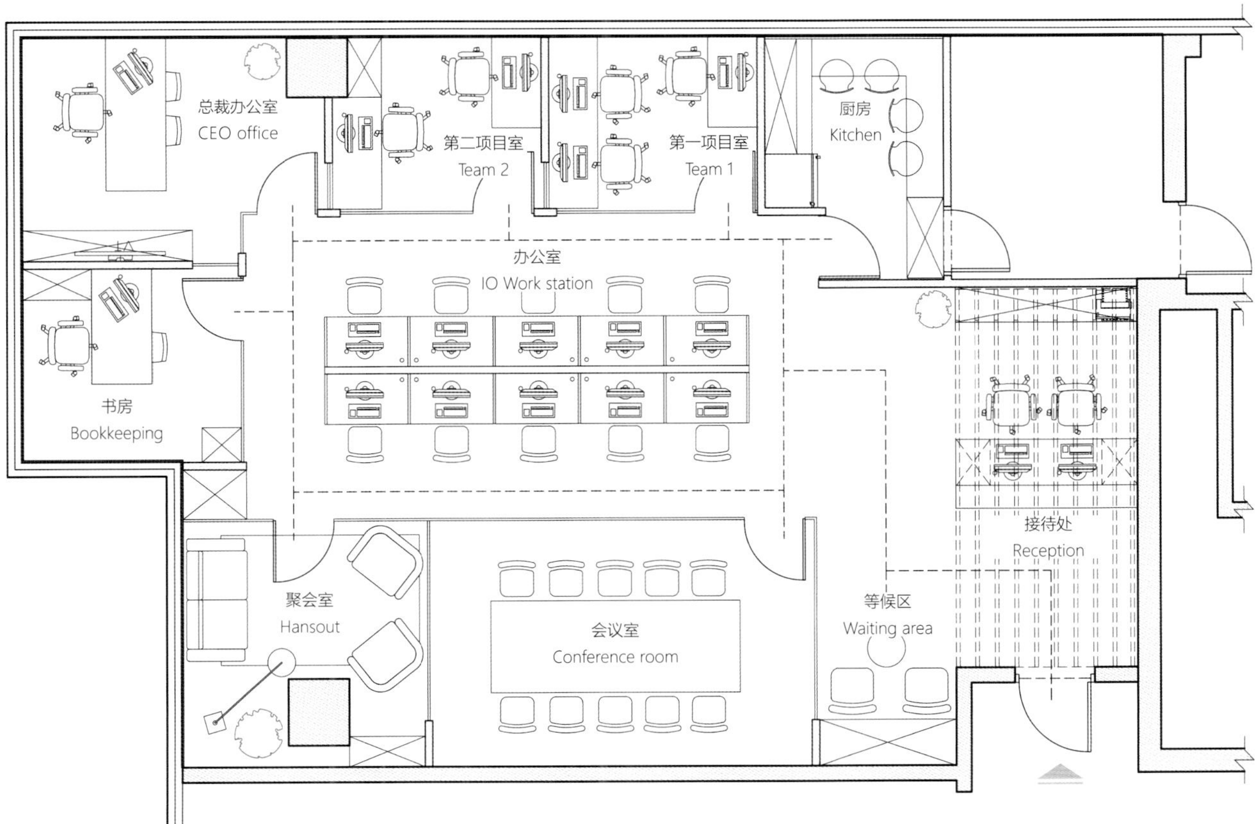

平面图
Floor Plan

BLANCO

办公空间利用天然木材和特殊的颜色搭配装饰，这些颜色不仅在空间的整体布局中凸显，而且还可以在位于空间入口处的小型装置中一眼看到。这种设计体现出生长的概念，并用沿着墙壁装饰的植物，以及从接待台衍生出的一系列木条作为象征物。

The office space is adorned with natural wood and a specific color palette, the colors are eminent not only in the overall layout of the space, but can also be spotted in a small installation situated right at the entrance to the space. The installation reflects on the idea of growth, symbolized by the use of plants along the walls and ceiling, with a series of wooden strips that stem from the main reception desk.

这个办公室的独特设计反映了建筑师的设计思维，其中包括鼓励使用矿物材料，使用基础结构和系统设施作为装饰元素，并不断探索空间的本质，以使其符合公司的文化。

The unique design used to create this office, is a reflection of the architects' design agenda that among other things encourages the use of mineral materials, using the structures infrastructure and systems as decorative elements, and a constant search of the essence of a space and making it fit the company culture.

DELL

Windward 办公室

Windward Office

项目地点：以色列，特拉维夫

项目面积：1200m^2

设计公司：ROY DAVID STUDIO

Location: Tel Aviv, Israel

Area: 1200m^2

Design Studio: ROY DAVID STUDIO

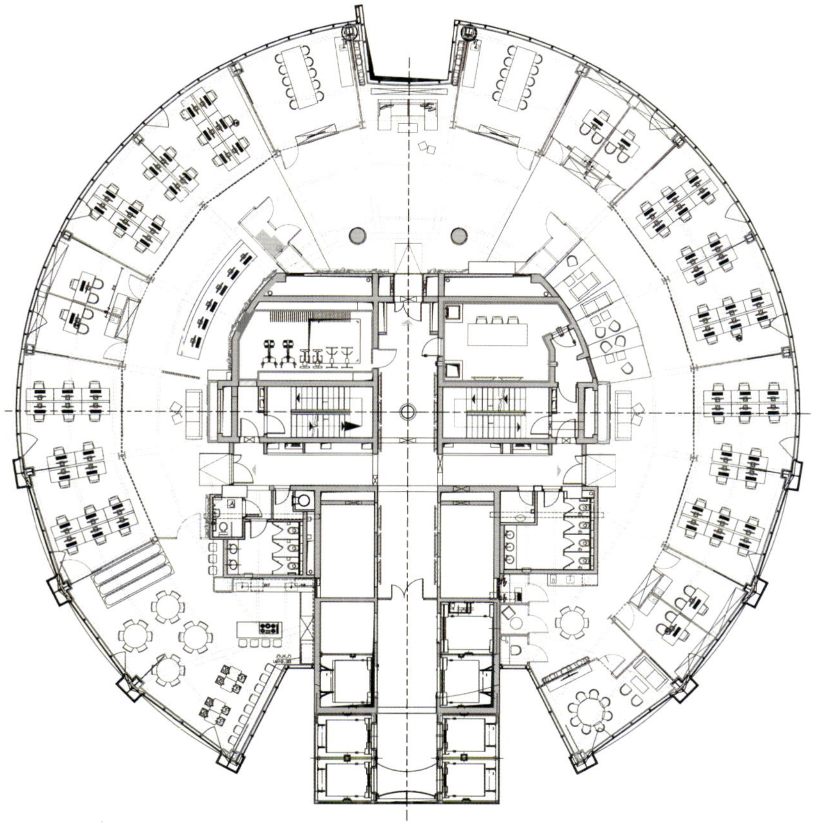

设计公司强调的是整体工业风格的塑造，以粗糙的原木、不加修饰的混凝土墙体以及锈色的金属结构凸显办公空间的工业气息。

The overall industrial style that the architecture firm focused on is emphasized by the use of rough wood, untreated concrete core walls and bold rusted metalwork as main materials for the new headquarters.

平面图
Floor Plan

这个空间巧妙地模仿了 Windward 公司的活动环境，有意避免了能够让人联想起豪华游艇的任何元素，而是渲染出一种如同工业港口般的粗糙氛围。这种理念扎根于每个细节，贯穿了空间设计的始终。

The space is intelligently designed so that it mimics the environment the Windward company undergoes its activity in. That the space would not reflect a luxurious yacht design but that of the harsh industrial ports. The design strategy is well rooted in the planning of the space.

环形的空间格局带来了不少挑战，业主还对空间的开放性与通透性做出了明确的要求，为了满足业主的需求，无论是开放式的隔音屏障，抑或 14 根锈蚀钢梁结构，都是针对设计需求量身定制的。港口般的工业氛围与高端定制办公家具之间的鲜明对比，让这个项目独树一帜，通过使用粗糙的数控设备金属元件和量身定制的通透隔墙、家具以及灯具，最终呈现出的空间氛围与 Windward 公司的价值取向和设计公司的设计初衷相一致。

Having a circular layout, the space presented a lot of challenges. company wanted their layout to be open space. From specially calculated open acoustic barriers to a supra-structure of 14 new rusted metal columns, each design element was added so that the companies brief was met fully. It's this contrast between an industrial port atmosphere and high-end tailored made office furniture elements that sets the space apart from any other project we did. With the use of CNC rough metal elements, custom designed partitions, furniture and light fixtures the space

is designed in a coherent language dictated by the core values of the Windward company and the architecture studio's vision.

“灵感 UM”办公室

“Inspired UM” Office

项目地点：立陶宛，维尔纽斯

项目面积：430m²

设计公司：PRUSTA LTD ARCHITECTURE & CONSTRUCTION

设计师：Ieva Prunskaitė, Rokas Puzinas

摄影：Leonas Garbaciauskas

Location: Vilnius, Lithuania

Area: 430m²

Design Studio: PRUSTA LTD ARCHITECTURE & CONSTRUCTION

Designer: Ieva Prunskaitė, Rokas Puzinas

Photographer: Leonas Garbaciauskas

办公室内部由天然未涂漆的镀锌金属主导，以镂空隔板进行分区，墙壁上有裸露的旧砖和图画。“灵感 UM”机构新办公室的主要概念是创造一个空间，从工作原理开始，以墙壁上的人物图像为开端，创造出惊喜，创造出想法，甚至放松。

The interior of the office is dominated by natural unpainted galvanised metal, openwork partitions graphically compose into the interior of the building along with old bricks and drawings on the walls. The main concept of the new office of "Inspired UM" agency was to create a space in which everything, beginning with the principles of work and ending with the drawings on the walls of respective personalities, inspired to create, surprise, generate ideas, and even relax.

一层平面图
First Floor Plan

作为图形细节的镀锌金属丝状隔板，也被视为整个内饰的结构元素，甚至成为决定意识形态和美学的起点。

Galvanised metal filigree partitions have been chosen as the graphical details and also as the constructional elements for this interior, even more, they have become ideological and aesthetic starting point for further decisions.

设计师用现代元素来装饰老房子：色彩层次丰富的旧砖墙、科技感十足的镀锌金属材质和温暖的木材结合在一起，甚至是工程系统在这里都具有装饰作用。一楼的木制天花板让人感觉舒适而温暖，天花板之间的间隙起吸音降噪的作用。

Designers tried just to enrich the current spirit of the old house with the modern elements: the multicoloured old bricks were combined with the laconic metal and warm wood inlays. Even engineering systems have got a decorative role here. Wooden ceilings on the first floor provide comfort and warmth, and the specific gaps between the ceiling boards were designed to absorb sound.

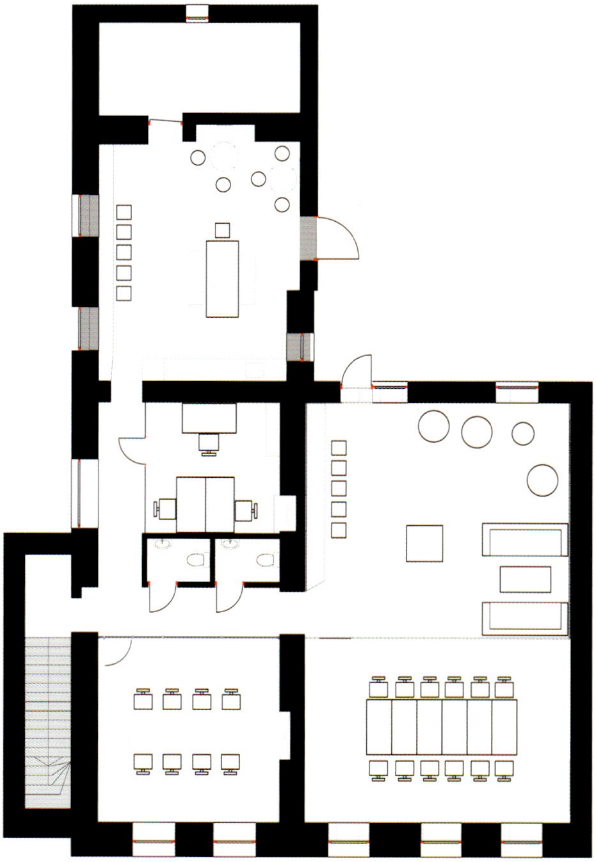

二层平面图

Second Floor Plan

Fiverr 以色列办公室

Fiverr Israel Office

项目地点：以色列，特拉维夫

项目面积：5000m²

设计公司：Setter Architects

设计师：ShirliZamir, Associate

摄影：Amit Gosher

Location: Tel Aviv, Israel

Area: 5000m²

Design Studio: Setter Architects

Designer: ShirliZamir, Associate

Photographer: Amit Gosher

Fiverr 的新办公室位于特拉维夫的 Beit Ha'ikarim，这是一座被列入保护名录的历史建筑，其中四层是旧建筑，三层是后来增建的。自从办公室所在的建筑物，包括新建楼层被挂牌保护后，一场介于新与旧之间、自然与真实之间的对话，被反映在室内设计之中。通过在特拉维夫历史建筑里发现的古老的纹理，来表现这个充满活力和国际化的建筑。例如，裸露的硅酸盐块旧墙与现代未经加工的建筑材料混凝土。

Fiverr's new offices are located in Tel Aviv's Beit Ha'ikarim-a historic, 4-storey buildinglisted for conservation to which three new floors have been added. Since the offices are located in a building listed for conservation, whose renovation included the addition of new floors, it allows a natural and authentic dialogue between old and new that's reflected in the interior design. The design produced representations of the dynamic and cosmopolitan city throughout its history, by relating to textures we found in historical strata of construction in Tel Aviv. For example, exposed old wall constructed from silicate blocks, alongside contemporary, unprocessed construction materials like exposed concrete blocks.

在办公室中，设计师把从 Beit Ha'ikarim 旧建筑中回收的木质窗框改造，作为展台和会议室之间舒适有趣的隔墙。普通的模块式建筑构件，用自然的形式摆放，创造出一个色彩丰富、有质感的三维墙壁，并将苔绿色色调引入到全新的办公空间中。同样，旧家具与现代家具也混合在一起摆放。

In the office spaces we used recycled wooden window-frames from Tel Aviv's buildings dating back to the Beit Ha' ikarim period. After modified their original purpose, they became inner partitions between the cozy interesting booths and conference rooms. Regular building-blocks are used in their natural form, creating walls with a three-dimensional, colorful texture, which introduces moss-green tones into brand-new spaces. Similarly, flea-market furniture is mixed with contemporary furniture.

Fiverr 公司通过其办公室的设计将国际艺术家和创意人士联系起来。所以设计整合了来自 Fiverr 设计团队为该项目制作的艺术作品。它们被并入到家具、图形元素和展示在混凝土墙壁上的艺术品中。

Fiverr wanted to connect the design of its offices to its international community of artists and creative people. And so the design integrates artworks produced by artists from the Fiverr community especially for the project. They are incorporated into the furniture, graphic elements, and the art displayed on the building's concrete walls.

一到六层是办公区域，员工在开放的空间工作，其中还有类似“电话间”的功能室镶嵌在其中。顶层有一个开放式的阳台，这是一个“公共”的区域，带有会议室、咖啡室和健身室，这些都是员工们的福利。

Floors1-6 are office floors. People work in open space, in which different kinds of conference rooms are embedded, like 'phone booths'. The top floor, with an outdoor terrace, is the 'public' area, with meeting rooms, a cafeteria, and a gym for workers' wellbeing.

Houston

LINQ 办公空间

LINQ's Office Space

项目地点：美国，马里兰州，巴尔的摩市

项目面积：370m²

设计公司：Turner Development Group

设计师：NFD

摄影：Jennifer Hughes

Location: Baltimore, MD, USA

Area: 370m²

Design Studio: Turner Development Group

Designer: NFD

Photographer: Jennifer Hughes

本案位于美国马里兰州巴尔的摩市，是 LINQ 电信管理公司总部的办公室。办公室的前身是世界上最大的谷仓，后来被改造成了豪华公寓，其中零售商店和办公空间在一楼。LINQ 办公室占据着大楼里面最大的商业空间，面积大约是 370m²。

设计公司对 LINQ 办公室的设计构想是脱离典型的办公空间设计形态，让设计更能反映出工作环境的创造性和协作性。设有整面落地玻璃窗的开放式布局，让灯具和台球桌融为一体，灯具的设计灵感来源于回收的旧木材。

The new corporate headquarters for the telecommunications management company LINQ, located in Baltimore, Maryland. LINQ's office space build out is located in what was previously the world's largest grain elevator that has since been converted into luxury condominiums, with retail and office spaces located on the ground floor. LINQ fills one of the largest commercial spaces in the building, occupying approximately 370m².

Turner Development's concept for "LINQ" was to shy away from the typical office space build out and make the design reflect more of a creative and collaborative working environment. The open layout with floor to ceiling windows was an ideal space to incorporate unique lighting fixtures as well as a pool table, with the design inspiration being reclaimed wood.

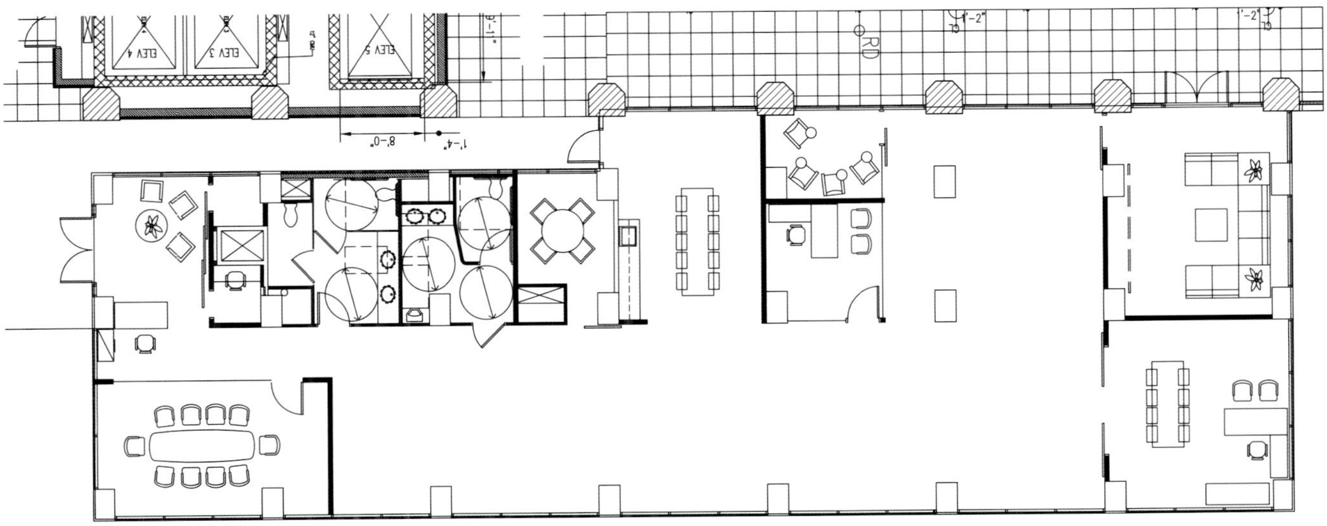

平面图
Floor Plan

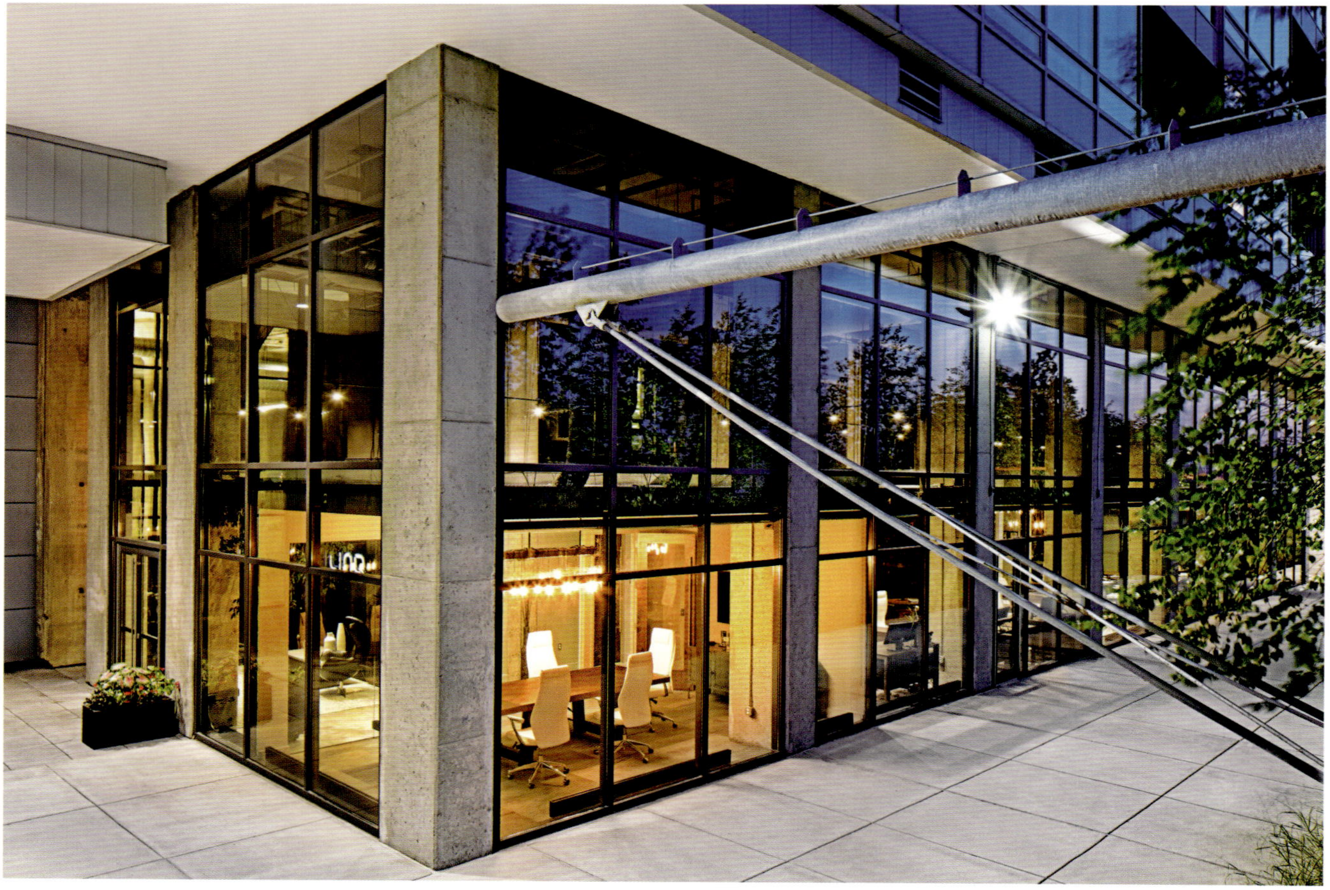
LINQ

LINQ

奇拓室内装修设计有限公司

http://www.chitorch.com/

奇拓室内装修设计有限公司，一个令人感动的空间，必定有着贴近内心深处的“独有故事”元素。通过量身订做来完成空间的独特故事性，并利用“冲突”营造不同美感。

CHI-TORCH Interior Design believes that a touching space is bound to possess unique elements of stories pressing close to people's hearts. It can be customized to accomplish story-telling feature and glamour with visual impacting.

DELORDINAIRE

http://www.delordinaire.com/

DELORDINAIRE 是一个建筑工作室，可以为客户量身定制室内设计和产品，总部在巴黎，并在蒙特利尔设有分公司。

DELORDINAIRE is an architecture studio for buildings, tailor-made interiors and custom design pieces, based in Paris with a satellite office in Montreal.

Egue y Seta

http://www.egueyseta.com/

Egue y Seta 是一个由室内设计、图形设计和工业设计等多个部门组合而成的设计工作室。为客户提供全方位的服务，始终将功能和预算一起考虑，而不是单纯的塑造美感。

Egue y Seta is a design studio founded by a multidisciplinary group of interior, graphic and industrial designers. It provides a wholistic service to clients that go beyond mere aesthetical concern bearing always in mind functionality and budget considerations.

Francisco Pardo Arquirecto

http://www.franciscopardo.mx/

Francisco Pardo 的工作转向了更加本地化和更加社会化的建筑设计，并把这个学科作为国家和社会发展的工具。他目前正在研究墨西哥城的 San Juan Pugibet 市场、普埃布拉州的一所农村学校、以及墨西哥的社会住房和各种回收项目。

Francisco Pardo has redirected his efforts toward more local and more social architecture, understanding this discipline as a tool for social development of his country and the world. He's currently working on the project San Juan Pugibet market in Mexico City, a rural school in Puebla, and social housing and various recycling projects in Mexico.

Haldane Martin Iconic Design

http://www.haldanemartin.co.za/

南非设计师 Haldane Martin 于 2002 年创立了他的同名设计工作室，以创造标志性的家具设计和室内空间，并具有强烈的认同感和概念深度。Haldane Martin 与商业品牌携手合作，将设计概念转化为具有独特个性的创新室内设计。

South African designer Haldane Martin founded his eponymous design studio in 2002 to create iconic furniture designs and later interior spaces with a powerful sense of identity and conceptual depth.Haldane Martin works together with commercial brands to capture the essence of a brand transform into innovative interiors with unique character.

Josephine Hurley Architecture

http://www.josephinehurley.com/

Josephine Hurley Architecture 总部在悉尼，提供建筑和室内设计服务。我们创造优雅、周到和永恒的空间，促进和提高日常生活水平。

Josephine Hurley Architecture is a Sydney based practice offering architectural and interior design services. We create elegant, thoughtful and timeless spaces that facilitate and enhance daily life.

Juan Felipe Goldstein 设计公司

https://www.juan-felipegoldsteindesign.carbonmade.com/

Juan Felipe 的创作灵感来源于他的南美童年时期的当代乡土建筑，他的目标是创造一种在保持环境完整性的同时改变空间的建筑。他努力创造开拓性的环境，同时尽量满足客户的需求。

Deeply inspired by the contemporary, vernacular architecture of his South American childhood, Juan Felipe's goal is to create an architecture that transforms space while maintaining its integrity with the environment. He strives to create environments that nurture and elevate while carefully meeting the needs of his clients.

Keivani Architects & Persian Primavera

http://www.keivani-architects.com/

Keivani Architects & Persian Primavera 由两位建筑师兄弟“Nima Keivani”和“Sina Keivani”于 2009 年成立。他们设计的项目涉及住宅综合体，商业建筑和建筑遗产修复等各种建筑。设计团队依托自己的理论和实践经验，运用最新的方法和技术，实现以社会、环境和经济可持续性为核心的创意理念。

Keivani Architects & Persian Primavera was established in 2009 by two architect brothers "Nima Keivani" and "Sina Keivani". Their designed Projects have been involved in various ranges of Architecture such as residential complexes, commercial buildings and restoration of architectural Heritage. Relying on their theoretical and practical experiences, the design team utilizes up-to-dated methods and technologies to actualize the generated ideas, in which social, environmental and economic sustainability is in the core.

Lien Tran Interior Design

http://www.lien-tran.com/

室内设计师 Lien Tran 于 1975 年出生于越南，她受到西方文化的强烈影响，但仍然在她的亚洲传统中找到灵感。

Interior designer Lien Tran was born in 1975 in Vietnam, she was strongly influenced by Western culture, but still finds inspiration in her Asian heritage.

MATT WOODS DESIGN

http://www.killingmattwoods.com/

mimaristudio 工作室

http://www.mimaristudio.com/

mimaristudio 工作室于 2006 年由 Onder KUL 和 Ayca AKKAYA KUL 创立，自 1995 年以来在各个著名建筑公司中积累了丰富的经验。

mimaristudio was established in 2006 by Onder KUL and Ayca AKKAYA KUL, with the accumulation of vast experience gained in various prominent architecture firms since 1995.

mode:lina ™

http://www.modelina-architekci.com/

mode:lina™ 于 2009 年由 PawełGarus 和 JerzyWoźniak 在波兹南成立。两位建筑师都是鹿特丹 Liong Lie / 123DV 建筑工作室的一员。

mode:lina™ was founded in Poznań in 2009 by Paweł Garus and Jerzy Woźniak. Both architects were part of Liong Lie / 123DV architectural studio in Rotterdam.

Nefa Architects (Nefaresearch)

http://nefa-architects.com/

PAB Architects

http://www.pab.com.tr/

PAB Architects 由 Pınar Gökbayrak，Ali Eray 和 BurçinYıldırım 于 2007 年在伊斯坦布尔成立。由于合作伙伴的背景不同，PAB 曾在多个领域工作，如建筑设计项目、施工现场实践、各种研究项目和建筑出版物。

PAB Architects is founded by Pınar Gökbayrak, Ali Eray and Burçin Yıldırım in Istanbul in 2007. With the partners'varying backgrounds, PAB has worked in a number of fields like architectural design projects, construction site practices, various research projects and architectural publications.

plajer & franz studio gbr

http://www.plajer-franz.de/

plajer & franz studio gbr 是一个国际性的智能品牌架构机构，为零售环境、百货公司、办公和休闲空间提供设计理念、室内设计和建筑设计。

plajer & franz studio gbr is an internationally active agency for intelligent brand architecture providing design concepts, interior design and architecture for retail environments, department stores, office and hospitality spaces.

PRUSTA LTD ARCHITECTURE & CONSTRUCTION

http://www.prusta.lt/

PRUSTA LTD ARCHITECTURE & CONSTRUCTION 自 1988 年以来一直在发展中。目前的项目包括在不同规模和预算下设计房屋、商业空间、办公室、景观、装修和室内空间。建筑活动领域包括从基础到关键的各种建筑工程、房屋建筑、立面改造、室内环境安装、施工维护和监理。

PRUSTA Ltd ARCHITECTURE & CONSTRUCTION company has been in practice since 1988. Current projects include houses, commercial spaces, offices, landscapes, renovations and interiors at many scales and budgets. Constructions fields of activity include all kind of construction works from the foundation to the key – f.ex., houses construction, facade renovation, interior and environmental installation, construction maintenance and supervision.

ROY DAVID STUDIO

https://www.roydavidstudio.com/

ROY DAVID STUDIO 是位于以色列特拉维夫的建筑、设计、概念驱动设计室。该工作室将数码艺术、景观和建筑融为一体，打造引人入胜的、发人深省的空间，为设计带来体验性和整体性。

ROY DAVID STUDIO is an architecture, design and concept-driven practice based in Tel Aviv, Israel. The studio offers an experiential and holistic approach to design by integrating digital art, landscape and architecture to create compelling and thought-provoking spaces.

Setter Architects

http://www.setter.co.il/

Setter Architects 成立于 1984 年，由 Michael Michi Setter 创立。Setter Architects 被视为以色列领先的室内设计公司之一，为公司以及各种私人和公共组织提供服务。该公司专注于鼓舞人心的、有创意和生产力的设计，强调将国际风格与客户独特的组织文化相结合。

Setter Architects was established in 1984 by Michael Michi Setter. Setter Architects is considered one of the leading interior design firms in Israel, servicing corporations as well as various private and public organizations. The firm specializes in inspiring, creative and productive design with an emphasis on integrating international style with the client's unique organizational culture.

Modijefsky 工作室

http://www.studiomodijefsky.nl/

Modijefsky 工作室是一间位于阿姆斯特丹的室内设计工作室，致力于荷兰和海外的多元化室内设计。这家由 Esther Stam 于 2009 年创立的工作室，以创意和定制设计为基础。工作室包括了由建筑师和室内设计师组成的国际团队，设计范围从酒吧、餐馆、酒店到公共场所，甚至是巴士设计。

Studio Modijefsky is an Amsterdam based interior design studio, working on a diverse portfolio of interiors in the Netherlands and abroad. Founded by Esther Stam in 2009, it is a boutique studio which hinges on creativity and bespoke design. An international team of architects and interior designers work on interiors and much more; from bars and restaurants to hotels, public spaces and even bus design.

Turner Development Group

http://www.turnerdevelopment.com/home/#index

Turner Development Group 是一个屡获殊荣的房地产开发公司，他们会将不同规模的项目"放大"。比如将一个废弃的谷物升降机变成海滨豪华公寓，将历史性地标剧院变成灵活的办公空间。

Turner Development Group is a multi-award winning real estate development company that think BIG on projects large and small. Big like an abandon grain elevator turned waterfront luxury condos and historic landmark theater turned sleek office space.

图书在版编目（CIP）数据

室内设计风格详解. 工业风 / 凤凰空间·华南编辑部编. -- 南京 ：江苏凤凰科学技术出版社，2018.3
ISBN 978-7-5537-8994-1

Ⅰ. ①室… Ⅱ. ①凤… Ⅲ. ①室内装饰设计－图集
Ⅳ. ①TU238-64

中国版本图书馆CIP数据核字(2018)第020449号

室内设计风格详解——工业风

编　　者	凤凰空间·华南编辑部
项目策划	宋　君　刘紫君
责任编辑	刘屹立　赵　研
特约编辑	韩　璇

出版发行	江苏凤凰科学技术出版社
出版社地址	南京市湖南路1号A楼，邮编：210009
出版社网址	http：//www.pspress.cn
总　经　销	天津凤凰空间文化传媒有限公司
总经销网址	http：//www.ifengspace.cn
印　　刷	广东省博罗县园洲勤达印务有限公司

开　　本	889 mm×1 194 mm　1／16
印　　张	17
字　　数	136 000
版　　次	2018年3月第1版
印　　次	2018年3月第1次印刷

标准书号	ISBN 978-7-5537-8994-1
定　　价	288.00元（精）

图书如有印装质量问题，可随时向销售部调换（电话：022-87893668）。